GREY SEAL, COMMON SEAL
A 'Survival' Book

Grey Seal, Common Seal

An account of the life histories of British seals

R. M. LOCKLEY

A SURVIVAL BOOK

Editor Colin Willock

WHITE LION PUBLISHERS LIMITED
London, Sydney and Toronto

First published in Great Britain 1966
by Andre Deutsch Ltd.

Editor Colin Willock

White Lion edition 1977

ISBN 7274 0122 X

Made and printed in Great Britain
for White Lion Publishers Limited.,
138 Park Lane, London, W1Y 3DD
by Hendington Limited,
Lion House, North Town, Aldershot, Hampshire.

Survival Books are published in close association with Anglia Television's Natural History Unit who make the 'Survival' series of television documentaries on wildlife.

THE SEALS

Leave her alone,
She is the Island's daughter.
Sleek heads, dark heads
Are risen from the water:
Leave her the company
Her songs have brought her.

The old grey music doctors
Of the ocean,
Their holy, happy eyes
Shining devotion,
Applaud and blow
In foam and soft commotion.

It is her hour.
The Island's only daughter.
The dark, sleek heads
Are risen from the water:
Leave her the company
Her songs have brought her.

L. A. G. STRONG

FOR FRANCES WHO HELPED

Foreword

As a keen angler I have never grudged the seals the odd salmon that might otherwise have fallen for my lure. But then I don't make my living out of catching salmon. I imagine that if I did so, I might feel differently about the patrols of Roman-nosed old men who bob in the surf just off the bar in the estuary of my favourite salmon river. The commercial netsmen have no illusions about the seals or what they are doing out there. According to them, the seals dine exclusively off salmon and sea trout and rarely look at bass, mackerel or flat fish. The netters shake their fists at the seal pickets and swear they wish every last flipper sunk without trace.

Not being a scientist any more than I am a commercial salmon netter, I don't know the real truth of the matter. I suspect that like most real truths it is halfway between the truths so rigidly believed in by the extremists on either side: by those who think seals are wonderful and those who see them as a menace to their livelihood. Seals being . . . well, I nearly said human for that's what they often look like . . . seals being seals, they undoubtedly relish a nice red middle-cut of fresh-run grilse. But then there are other fish and creatures in the sea, many of which we landlubbers don't prize nearly so highly as salmon. The seals surely eat these in fair quantity as well.

On one part of the coast, at least, the salmon men have partially got their way. The cull of grey seal pups that is taking place on the Farne Islands as this book goes to press will have done one thing, even if it does not turn out to have increased the harvest of the inshore salmon fishers. It has revealed the strange interest and fascination that seals hold for a race of people who have the sea, however diluted by modern living, in their blood. As in the case of Goldie, the golden eagle who escaped twice from the London Zoo, there is little doubt where public sympathy lies. It lies, however misguidedly, with the creatures playing tip and run with Authority. As November comes round and the seal cullers load their humane killers, you can almost hear both newspapers and their readers praying that the weather will cut up nasty so that the boats can't get out to the Farnes while the pups are still helpless ashore. The newspapers, of course, are interested in the story. Their readers are interested in the fate of the seals.

Understandably so. Seals are extraordinary and fascinating creatures. They are, for the record, Britain's largest mammal. Not even a red stag can beat them for size. We have seals, both common and grey, around our shores in astonishing numbers. Even though the sea is their domain they are finding themselves caught in that inevitable confrontation . . . man versus wildlife. As such they are a worthy subject for this Survival series of books.

When we decided early on to plan a book on seals, only one author seemed possible. Whenever I think of Ronald Lockley I have a memory of him rushing me across a cliff top in Cardigan with the promise of one of the most staggering sights I could hope to see. I wasn't disappointed. After a long gallop we arrived at a most secret and secluded bay. We stood three or four hundred feet above it on the lip of a nearly vertical slope. Below us, on a tiny crescent of rock and sand, lay upwards of two hundred grey seals. Lockley had shown me his Red Wilderness. It was in such places that he did much of his work on grey seals. He tells the story of this work in this book. Ron Lockley is not only a great field naturalist. He has the great story teller's gift of making you live his experiences and feel his enjoyment.

COLIN WILLOCK

Acknowledgements

My warm thanks are due to the author and the publishers of the poem 'The Seals' by L. A. G. Strong, published by Messrs Methuen & Co Ltd, for permission to reprint this delightful tribute to the seals.

And to many fellow observers of seals, both those who have accompanied me on expeditions, and those who have talked with me about seals and written about them in books and other publications: I have tried to acknowledge these friends by name in the book.

A full bibliography of all the works consulted would occupy too many pages, but I am particularly grateful to certain authors who have shed light on the lives of the seals: R. Carrick, F. Darling, H. D. Fisher, R. J. Harrison, H. R. Hewer, Grace Hickling, S. E. Ingham, A. L. Johnson, G. L. Kooyman, R. M. Laws, L. H. Matthews, V. Scheffer, A. Soper, and L. S. V. and U. Venables.

The Nature Conservancy supported some of the work I carried out in marking and counting seals in Wales and Ireland, and in this task I received the encouragement of two of its officers, F. B. Worthington and E. A. Smith. As honorary warden of the West Wales Naturalists' Trust I directed the marking of seals in Wales on behalf of that Trust, grant-aided by the Conservancy.

R. M. LOCKLEY

Contents

	Foreword	*page* 7
	Acknowledgements	9
	Introduction	15
1	On the Intelligence of Seals	19
2	Calf Seal	31
3	The First Year	47
4	Adolescence	54
5	The Seals in Summer	61
6	The Life Aquatic	72
7	The Breeding Season	85
8	The Common Seal at Home	114
9	Population and Vital Statistics	118
10	Seals and Men	128
11	Distribution and Numbers	149
12	Predators and Mortality	166
	Ballad of Sule Skerrie	171
	Index	174

Illustrations

Between pages 80-81

1 Common or Harbour Seal (*P. Morris*)

2 Grey seals mating (*R.M. Lockley*)
Grey seal suckling her calf (*Barrie Thomas*)

3 Male and female common seals photographed together under water (*F. Schensky*)

4 Common Seal underwater (*Anthony & Elizabeth Bomford*)

5 Face of a grey seal calf (*Barrie Thomas*)

Grey seal calf which lost its mother at birth
The body shrivels up during the week in which it may survive without food (*R.M. Lockley*)

6 Diana, an orphan common seal reared at Orielton, adored my secretary (*R.M. Lockley*)

Clipping a numbered ring on a three-week-old grey seal (*R.M. Lockley*)

7 Grey seal nursery, Orkney (*R.M. Lockley*)
Grey seals on the beach in the Red Wilderness (*Barrie Thomas*)

8 Master bull in the centre of his harem: an inland colony, Orkney (*R.M. Lockley*)

Panic at an inland colony of grey seals (*R.M. Lockley*)

9 Grey seal (*Anthony & Elizabeth Bomford*)

10 The Red Wilderness from the cliff top (*Barrie Thomas*)

11 Grey seal bull and cow (*R.M. Lockley*)
Activity off a grey seal nursery beach (*R.M. Lockley*)

12 Common Harbour Seal (*K.W. Fink*)
Common Seal underwater (*Anthony & Elizabeth Bomford*)

TEXT FIGURES:

1 Annual breeding cycle 62

2 Mode of progress on land 74

3 Skeleton of a common seal 76

4 A Welsh nursery beach 87

5 Part of an inland colony, Orkney 99

6 World distribution of grey seals 151

7 World range of common seal 153

8 Main colonies of grey seals in British Isles 154

9 Recoveries of marked grey seals 159

10 Seal net 161

Introduction

Seals are warm-blooded creatures like ourselves. British seals have calm soulful eyes, larger than those of man, with which they gaze forwards with a concentrated stare, the rich dark pupil and iris filling the aperture with little white showing. This gives the head a warm, kindly, human look, the long whiskers adding a wise adult touch. It is not at all surprising that their faces and ways have given rise to so many legends and tales of mermaids and mermen.

In fact seals are sentient beings of an advanced order of intelligence not far removed from that of man, ape, dog and dolphin. In one respect they have advanced farther from the ancestor which they share with these mammals. The structure of the skeleton and limbs of the seals show that they once lived terrestrially with four articulate limbs able, like the polar bear and otter today, to walk about on dry land. Having become land mammals with man, they are returning, have almost completely returned, to the sea – which may yet be man's not too distant fate when the land fails to satisfy his requirements and if he continues to destroy his environment from lack of foresight.

Yet seals are not wholly marine, like the whale and the dolphin. They have not lost their covering of fur like those mammals in which only a few vestigial hairs remain. Seals still like to get out of the water and lie in the wind and sun for long hours, and rest and sleep. They still bear their young on the land, or on the ice, although the young of some species is able to swim at birth if by chance it should happen to be born in the sea; in other species the young seal will drown at birth unless brought quickly ashore.

The seals have been classified by their structure in the Order *Pinnepedia* (fin-footed), which is divided into three families: the *Otariidae* or eared seals such as fur seals and sea-lions, having long limbs and flexible hind flippers which can be turned forwards, enabling them to walk or shuffle on four feet partly upright on land, and they have well-defined external ears; the *Odoboenidae* or walruses, well known for their long ivory tusks; and the *Phocidae* or true seals, which have trailing hind limbs partly encased in the flesh of the trunk and useless for locomotion on land, and small, inconspicuous retractable ears.

This book will describe principally the lives of the two British breeding seals, which belong to the last family. The larger species is the grey or Atlantic seal, *Halichoerus grypus*; from tip of nose to tip of hind toe adult males are about seven feet long and weigh about 500 lb, but they can be nine feet long and twice as heavy in their prime. The smaller is the common or harbour seal (sometimes called the brown or sand seal), *Phoca vitulina*; adult males are about five feet in length and weigh 350 lb, but they can be six feet long and weigh up to 600 lb.

The common seal is a vigorous species, distributed over the whole of the temperate and sub-Arctic northern hemisphere seas, including the Arctic, Atlantic and Pacific oceans; and is divided into five subspecies which intergrade with their neighbours at the territorial limit of each. The minimum world population of *Phoca vitulina* has been estimated at 140,000 individuals; but it may be double.

The grey seal has no subspecies, although it has different breeding seasons in its several discrete territories in the British Isles, the Baltic, Norway, the Faeroe Islands and Iceland, and the eastern coasts of North America. The world population of this seal is quite low, about 50,000, of which nearly three-quarters breed along the shores of the British Isles. Such a figure (less than the attendance at a single Football Association match in Britain) show that *Halichoerus grypus* is one of the rarer mammals of the world and one of the rarest of the seals; and therefore deserving of special protection on our shores, its main stronghold, lest under modern trends of predation and other adverse and increasing pressures it becomes extinct.

The adult grey seal can be distinguished from the adult common seal by its larger size and heavier weight; but immatures are superficially very alike, and identification of these at any distance is quite difficult, for the brown, grey and black coat colours are similar. In profile the adult grey has a longer more hound-like face, with the nose blunt and the nostrils wide apart. By comparison the common seal has a smaller, more rounded head with almost a snub nose, the face in profile is slightly dished; front view the face is cat-like, the nostrils close together to form a distinctive V.

In this book we use certain terms which are in general use among students of seals, and have their origin in the terminology of the early seal hunters. Male and female are known as bull and cow. Logically the offspring of these should be known as calves, and many writers use this term. But the seal hunters of the Arctic and Antarctic have always called them pups – presumably from the facial resemblance, and this term has now been generally adopted in the scientific literature. We prefer the term calf, but shall use both here. The dominant breeding bull in a seal nursery (often called 'rookery') is known as the master bull, or simply 'beachmaster' (chiefly Antarctic herds). The mature unmated bull seeking to mate with mature cows is a challenger bull. Younger unmated bulls are known as bachelors. The mature female before her first mating is a virgin; and other terms used in this book are self-explanatory. Most seals are full-grown at four years of age, although they may not achieve fertile mating until later.

I On the Intelligence of Seals

'One day,' said Ronnie Pearson over the phone, 'I intend to ride my tame dolphin across the English Channel.'

'Yes, you said that in your letter,' I replied tranquilly, 'I have adjusted my thoughts on your unusual proposal. I have slept on it. I have decided that I do not intend to accompany you unless I am in a comfortable boat. Anyway you'll need someone around in a fast boat in case you fall off. Dolphins I imagine are swift and slippery steeds . . . What? You haven't got a dolphin? You mean you want me to be responsible for catching one for you? Well, they aren't too plentiful off the Pembrokeshire coast, but I could try. But it'll be expensive. How much? Say eight hours' hunt at sea in a fast launch at £3 per hour. Catching nets, rubber foam mattresses and gear extra . . . What? Your BBC subsidy would cover it? No? Then what do you suggest? Speak up. I can't hear you . . .'

The line was a bit faint, but I could at last hear Ronnie shouting the substance of what was something quite different into the phone. His dolphin project had already been broached in his last letter to me, and I had become tuned in to thoughts of how to help this imaginative friend, but it seemed that the plan had taken a new twist . . .

'Of course the dolphin plan is on,' said Ronnie, from 200 miles away, 'but just now I really wanted to ask . . . three new-born baby common seals . . . picked them up in the Wash, after a Min. of Ag. and Fish shooting orgy . . . Orphans, and all three are ladies . . . Hope to train them in my private lake . . . preliminary canter, I mean swimming trials . . . temporary substitute for dolphins and much easier . . . keep my hand in with training marine mammals . . . aqualung experience underwater with seals useful knowledge for Channel adventure with dolphins. But, Ronald Lockley, I'm having serious feeding problems with these seal babies . . . afraid the lot'll die. The heavenly trio not doing well, no gain in weight since I had them . . . diabetic trouble . . . expensive diet . . . all goes in diarrhoea. Can you take one as an insurance against total loss? I know you have reared orphan seals before. Do come up to Baynton and see them and advise me, please. I'm very worried.'

In the end Tony Soper of the BBC brought one of the three orphan common seal pups from Baynton, Ronnie's ancestral home and lake in Wiltshire, 200 miles by Land-Rover to Orielton. On a diet of fish meal mixed with milk and cod liver oil they had been doing badly. Also the healthiest of the trio had apparently been bullying the other two – which were soon to die, as feared.

When we saw the inquisitive cat face of the week-old seal leaning thoughtfully out of the side window of the Land-Rover we lost our hearts to her at once.

Unwrapping the highly odoriferous and insanitary sack-pants from her mermaid's tail we slid her into the fresh water of a large shallow concrete pond in front of Orielton House. She dived gratefully and like an arrow was soon speeding round the edge of the pond, a long lean fish-like huntress. We named her Diana forthwith. She lived happily, in and out of this pond, for two months.

Fed through rubber tube thrust forcibly down her throat well into her stomach, into which we poured, four times a day, a mixture of margarine (containing whale oil), Jersey full-cream milk and cod liver oil, the whole heated to blood temperature, Diana began to put on weight immediately. She even grew to like and to play with the rubber tube, unpleasant though the moment of its insertion was for her; but the sensation it soon brought, of a full stomach, had her approval.

As soon as possible, after three weeks of milk and oil feeding, we got her on to herrings, at first pushing them forcibly down her throat. Then one day – delightful sensation – she discovered she could suck in a small herring herself. Her appetite for them thereafter was almost insatiable.

She swelled rapidly and her coat glistened with good feeding on these oily fish. Living on our front doorstep Diana became a greatly loved friend to all except visiting dogs (she cordially hated them because of their aggressive, frightened barking). Even the cats venerated Diana, because of the smell (and scraps) of fish which she left around. She was adored by children, and she in turn loved the company of those who talked to her and gently stroked her greasy skin.

If any one of our household appeared at the front door Diana would leave the pond at once and shuffle up to meet us. She joined every party on the croquet lawn below the pond, inquisitively inspecting the tea-tray and if possible sprawling across it in search of fish. She quickly knew those who were kindest to her, and would wriggle up to sit beside them. If no fish was forthcoming, a gentle scratch or touch with your hand pleased her, and she would roll like a contented puppy at your feet.

Diana regularly interviewed visitors, and was expert at amusing them if we were not available to do the same immediately. Many an unsuspecting caller would be invited to sit in the sun on the stone bench by the pond to await the master. With twitching nose and dripping whiskers Diana's head would rise from the water. She would proceed to slide across the gravel and approach the astounded, alarmed or delighted spectator, who might smile, scream, or even flee altogether. But if he or she had the courage to sit out the interview with Diana, the little seal would probably end by basking at the visitor's feet.

Without being taught any circus tricks to test her brain capacity Diana, within two months of birth, revealed a degree of intelligence exceeding that of a house dog, or of an intelligent child, of as many years. What specially endeared her to us was her love of human companionship. Without being petted (for I had an idea that one day I would return her to the wild sea, and therefore she ought not to become too dependent on human society), she came wriggling on her tummy to us for love, seeking to rest her head on our lap if she could reach it, or on our feet if not. At all times she tried to follow us into the house, and if permitted would settle with us comfortably by the fire, or in an armchair (not – for obvious reasons – without a newspaper *under* her) and listen to our talk, head raised quizzically at first; but soon she would droop and doze, bored with our verbiage but happy in our company. At night she would start to climb the stone stairs to follow me to bed. But as we had never attempted to train her to be sanitary indoors (probably impossible in an open-air marine mammal), and as her fish diet gave her a powerful piscine odour, she would have to be put outside. She quickly learned to respect the command 'No!' and her look of reproach, fancied or real, was heartrending. Picked up in our arms she would wriggle and moan in protest for a moment, knowing she was going to be ejected, then would almost smile with delight at this temporarily pleasurable close contact.

She slept tight against the front door, and rolled into the house as soon as it was opened in the morning if one was not watchful to stop her with a loud 'No, Diana!' She understood this order perfectly, but always hoped we did not mean it. She knew that children might not be so heartless as to exclude her, and with them she would wedge her nose in quickly and force the door to stay open. They admired her so much, and would watch her slide over the doormat with glee.

Diana loved to ride in a car, especially if she could be with a group of children who were not so squeamish as adults about her slippery smelly fur. They would ask daily if they could take her to the sea. Her mermaid tail would be inserted in waterproof trousers for the trip, and she would be propped up on sack cushions so that she could look out with the children and see the countryside whirl by. If occasionally the children were too boisterous with her, she would nip them – her sharp little teeth soon taught them manners.

Given the freedom of a large rock pool on the shore she swam around with her human friends, and chased the little rock fish and the shrimps. But when they left her and went to sunbathe on the warm sands Diana would moan with protest and follow them, and loll with them inseparably in the drying sunlight. Even when we put her right into the open illimitable sea, which in Pembrokeshire stretches 3,000 unbroken miles to America, Diana returned to the land as soon as the other bathers did, and insisted on being at hand when the picnic basket was opened. At this stage complete freedom did not interest her. She still needed the affection and protection of the humans who had taken the place of her mother.

In her book *Seal Morning* Rowena Farre describes how her pet common seal would lay out the picnic on a cloth with the patience and skill of a trained waiter. I do not know how much, if at all, this is an exaggerated account of her pet's intelligence, but Diana, without any training whatsoever, would tackle the picnic basket in no uncertain fashion. But then all she exhibited was a very crude desire, like the bad manners of many young humans, to nose out her share of the picnic first – a herring wrapped in newspaper. She could smell it and she would dive straight for it, if not restrained and ordered – nay, pushed – aside. She would then sing her little moan of protest and hunger.

As for being musical, as Rowene Farre claimed her seal was, Diana showed a very low limit of appreciation, even of crude contempt, when a young woman offered to play seductive music on a flute and test Diana's ear. I am deeply ashamed to record this occasion, which took place in the lovely Regency drawing-room at Orielton before an invited audience; but this is a scientific attempt to evaluate the intelligence of the seal, and I ought not to hide the awful embarrassment Diana caused us on hearing the first soulful wail of the woodwind. She lifted her head sleepily from under the hand which had been stroking her, and which was now raised so as to give her space for reaction uninfluenced by love. The assembled company in breathless excitement now expected Diana to break into song, as did Miss Farre's seal.

Alas, Diana's only response was to eject upon her newspaper carpet a bubbling stream of farts and faeces. Then she dropped her head again in sleep.

Diana was not the slightest bit interested in music, although we played to her on the flute, recorder, guitar, violin, bassoon and piano. In fact it must be regretfully admitted, despite many assertions to the contrary and the many legends and pretty stories of singing seals, that although they have almost human voices, seals are not musical in the human sense. They have however acute hearing; wild seals will listen for a while to human-made seal-like noises – whistle, flute, woodwind, etc. – if these be played by an observer hidden near them, but once their intense curiosity is satisfied that the source of the music is of no threat to them they lose interest and resume their usual occupation; should the observer suddenly appear in full view they will normally dive and swim off. They do not respond by moaning or singing themselves. Seal conversation is strictly limited to communication between themselves, and they will snarl if cornered, threatened or assaulted by man.

Diana proved that common seals are intelligent. What about other seals? Sea-lions are notably easy to train, and seem to delight in learning clever, acrobatic feats – for suitable fish rewards. I had always assumed that some species of seals were more intelligent than others. Compared with the brain of the much larger grey seal, that of the common seal is, weight for weight of the adults, heavier. The common seal ought to be more intelligent than the grey; but is it? Our experience in rearing orphan seals suggested this, because we found it more difficult to make friends with the grey seal pup than we had with Diana. The grey seal pup at birth does not attack man, but within a day or so of being suckled in its natural wild state it seems already to have acquired its dam's mistrust of man, and will bite and hiss at the intruder. It can be induced to stop showing animosity if it is fondled and petted for half an hour or so when only a day or two old – it becomes bored and indifferent at that age. But it is difficult to train when taken into captivity at a later age and cannot be trusted not to bite even a loving guardian.

Sabrina was a grey seal pup a week old when she was found beside her mother who lay dead of a severe wound on a beach in north Pembrokeshire. The local RSPCA van arrived at Orielton one September morning, and dumped the sack-swaddled white-coated baby beside the pond.

'It's all yours,' said Inspector Smith with a sigh of relief.

Sabrina hissed and snapped at Diana and at us. She was in fair milk-fed condition, so her mother could not have been killed long. We tried her with warm milk in a pet-lamb bottle – in a rage she bit the rubber teat in two. With a rubber tube we gave her daily 6 lb of the milk-margarine-cod-liver-oil mixture force-fed deep into her tummy; but each time it was a severe struggle to hold her powerful jaws open and push the tube down. After feeding her we would slide Sabrina into the pond with Diana. But she ignored Diana or snapped at her friendly approaches. Diana gradually lost interest in her ill-tempered relative, and was indeed afraid of Sabrina.

At this age Sabrina disliked water – in great contrast with the non-stop bather Diana – but it was necessary to clean her face and tail after meals, and simplest to do this by slipping her into the pond. During her first week with us (her second week of existence) Sabrina would immediately climb out of the pond and retreat to the low railing around the croquet lawn – as far as she could get from human activity. She would lie there, basking in the sun and sleeping, aloof and unable to enjoy the love we offered. When visitors approached she lifted her head and growled warningly.

Sabrina had good cause to distrust man, who had slain her mother, who had pushed her around, fed her forcibly, and generally made himself a nuisance when all she desired was to be breast-fed when hungry and to sleep, alone, when sated. Later in this book we shall see that the grey seal pup, in nature, is hostile to all other animals except its mother; and that such hostility is valuable in securing living space in the nursery for the more dominant pups best fitted to carry on the species.

It would not be fair to say that Sabrina's refusal to know and love us, as Diana did, was an indication of a lower level of intelligence – although it might be. The grey seal can show a considerable degree of intelligence. H. G. Hurrell succoured a two-month-old grey seal which had been found in a battered condition after a severe November gale. He placed her in his outdoor swimming pool, nursed her back to health, fed her on herrings, and taught her quite a repertoire of tricks. She has survived to the day of writing this, five years later. In his book *Atlanta My Seal* Hurrell shows that Atlanta learned – for a reward of fish each time she complied – to obey by word of mouth at least two dozen distinct commands. She was even able to pick out and act upon these when the operative words were mentioned casually in the middle of a sentence spoken in conversation between Hurrell and myself – a remarkably sensitive ear she had. At two years of age she 'learned to read' simple monosyllables such as Down, Clap, Roll: that is, she dived, clapped or rolled when these words, printed large, in white on a black card, were shown to her. I have had the pleasure of asking her to wipe her nose, and received a courteous and instantaneous acquiescence.

Atlanta is much beloved by the Hurrells; but unlike Diana she has showed no disposition to be loving towards her guardians. She performed her tests accurately, not out of love but for the good reason that she was not rewarded with food until she had. Once full-fed she refused to obey any more commands, until she was hungry enough to work for a meal. She was full-fed, at four years of age, after her twentieth herring in as many hours; but this was gluttony – or satiation – her daily ration was a baker's dozen.

Sabrina began to swallow herrings at the age of three weeks. At that moment we initiated her into the art of catching fish, as we had Diana, by placing a live eel in the little pond in front of the house. Her reaction to a dead fish placed there had been nil, but the wriggling movement and flash of the live eel at once excited her, as by instinctive reaction, and she chased, nosed, and nibbled at the eel, played with it like a cat with a mouse, but did not immediately eat it. In a few more lessons, being now an expert swimmer, she learned to catch the more appetising oily dead herring, which we towed on a string jerkily under the surface of the water. So long as it moved she swam after it. Her expression we fancied, on first swallowing a juicy fresh herring was as surprised and beatific as Diana's had been. All seals love herring, which are rich in fats essential to blubber development.

With her skill in catching live eels and dead herrings well developed at a month old it was time for us to set free Sabrina. Our fish bill was assuming alarming totals each week. We released Sabrina, still resistant but well accustomed to the sight of man, in the sea under the steep cliffs of Pembrokeshire, where a gully gave access to the tide-line. She could not climb back from the sea easily, but she did not try. Without a backward glance at those who had saved her life by nursing her she began to hunt for fish among the Atlantic breakers. We had placed a serially numbered ring on one hind flipper, bearing the legend 'Inform London Zoo'.

A month later, on November 5th, she swam ashore in a northerly gale at Port Quin, North Cornwall, some seventy miles to the south. 'It came up,' reported the fisherman, H. Collings, who found her, 'about thirty yards above high water, and five or six hours later, when the gale calmed down, went off to sea again. It was tame enough for me to read the number and address on the ring on its hind-flipper.'

We were delighted that Sabrina had met with a kind fisherman in such circumstances. A year or so later we were to meet Mr Collings and hear his account of her visit direct. But Sabrina has not been reported since. No news, we hope, is good news. We trust that by now Sabrina may be a mother of many seasons.

Diana was not to be so fortunate in her return from sea to the land when she eventually became a free-living seal. As both her infant contemporaries had died at Baynton after Ronnie Pearson had contracted to feature them on television, he signalled me desperately for the temporary return of Diana to understudy them. We drove Diana, content and smelling of herrings but affectionate to a degree, in my daughter's arms, the 200 miles to Baynton. She became a TV star in both above and under water films taken by Tony Soper for the BBC. Ronnie taught her to tow a little boat in which he sat at the reins, proud but wobbly.

Much as we missed Diana on our doorstep we were next requested permission to allow her to undergo BBC trials at the underwater school at Dartmouth. All went well until Diana (as related in Ronnie Pearson's book *A Seal Flies By*), now six months old, became bored with underwater cameras and stale mackerel. Maybe she wanted to get away and find a quiet place in which to moult her birthday suit? We shall never know.

She swam up and down the Dart, so near to the wild free open ocean. She rejected the coos and calls and fish offerings of Messrs Tony Soper and Ronnie Pearson. For a while she revelled in the shoals of winter herrings of the estuarine sea, alone and free at last. She swam far out into the English Channel. Then, a few weeks later, she hauled out to sleep on a lonely strand in Lyme Regis Bay. There she was brutally murdered by some passing lout – that is, the mutilated body of a common seal of Diana's age was found. Diana had escaped before I had had a chance to mark her with a numbered ring, as I had Sabrina; so there was no positive evidence that the slain seal was her, but common seals are rarely or never found on this part of the English coast.

How anyone could kill a young common seal lying helpless, asleep, or just peacefully awake and prepared to be friendly, is difficult to imagine. To a civilised being the charming cat-like face of this seal is most appealing – to batter it to death is an action akin to infanticide.

Another young common seal fared a good deal better when it settled for the summer in a popular bathing cove at Cuckmere Haven, Sussex. It became well known for its fearless and friendly behaviour. It liked to swim among human bathers; if unmolested it would lie dozing on the rocks at low tide, admired by many visitors. Possibly, like Diana, it might have been an orphan, reared by some kindly person who had put it back in the sea when it was old enough to fish for itself, when perhaps the fishmonger's bill had become too heavy a burden? It is difficult to account for its tameness otherwise. It may be that some individual common seals are by nature more sociable than others. Adult common seals are known to become tame quite rapidly when taken into captivity and treated kindly.

Then there was Sammy, a yearling grey seal, who swam into a Dorset cove one summer, and made friends with people on the beach, coming ashore to bask among them, and delighting to be gently stroked. Nina Warner Hooke describes in *The Seal Summer* how Sammy loved to swim with the bathers and play games with those tough enough to stand his idea of fun, which included pulling them under water. There is no record of Sammy having escaped from captivity to account for his tameness; his friendly visit over several summer months appears to have been due to the adolescent seal's normal desire for company. Possibly Sammy had taken the wrong turning when migrating past Land's End on his way back to Wales from a winter in Brittany (where marked Welsh seal pups are known to visit), and instead of keeping north to the spring assemblies of his kind in Welsh waters, had veered east into the English Channel by mistake, and so reached Dorset. Here seals are rare, and so perhaps Sammy, seeking company, had found it in the beach assembly of human idlers. Mrs Hooke records that Sammy swam away (we hope to Wales) in the autumn, and left behind an indelible memory of a friendly, kindly seal soul of, for his tender age, exceedingly high IQ.

Of other marine mammals in a wild state, the vegetarian dugong and manatee of tropical and sub-tropical waters are said to be as unintelligent as they are ugly, although with their fish tails they are nearest to the mermaids and mermen of romantic legend; and it was the distant sight of them on rocks and strands that gave rise to many reports of human beings with fish tails living in the sea.

In contrast the families of the whales and dolphins, it is now discovered, have remarkably acute intelligence, as exhibited in captivity in large marine aquaria. They can be taught to perform all the tricks, and more, in water which are learned by performing sea-lions. Their brain capacity is large, their disposition when young, gentle, their love of play and games that of happy children. They have a considerable vocabulary of their own, whistles and clicks and other signals by which they communicate under water; and they use sonar or sound echoes to locate objects under water, including each other and their fish prey. An experimentally blinded dolphin can steer and fish entirely by echolocation, that is from the echoes, against objects at a distance, of sounds it utters as it swims about, easily avoiding objects deliberately placed in its path and which it cannot, of course, see visually. But then a blind seal can also live, feed and breed successfully (page 168).

In *A Book of Dolphins* Antony Alpers has gathered together both legend and true story of the friendship of dolphin with man. The account, for example, of Opo, the dolphin which frequented Opononi Beach, New Zealand, shows that this tame and gentle creature delighted to be stroked and fondled by those who waded out to her in the shallow water. She freely encouraged children to ride on her back. She liked to hear the laughter of visitors watching her antics, which included tossing a coloured beach ball into the air, catching it on her nose or hitting it with her tail, or trying to force it under her body. She was protected by a special Order in Committee (Fisheries (Dolphin Protection) Regulations, 1956) forbidding anyone to take or molest a dolphin in Hokianga Harbour, on pain of £50 fine. She was too tame, and coming close to a motor boat propellor, received a nasty wound; then one night, venturing too close inshore on a falling tide, she was stranded between rocks and died before she could be found. She was given a public funeral. Her affection for human beings and her delight in playing with them make Ronnie Pearson's plan to cross the Channel on dolphin back less fantastic than it at first sounded to us.

As far as we know seals do not have this vocabulary of sounds and underwater sonic-echo location to the extent discovered in whales and dolphins. But much research still needs to be done. All we know at present is that seals have good sight under water, fairly good vision above water, keen hearing even although the true seals have only small hidden retractable ears, and a sense of smell by which they recognise each other, or at least male from female, and own child from stranger child. Sounds uttered by seals seem to be limited to not unmusical groans and moans in the grey seal, the bull having a deeper note than the cow. The common seal pup has a soft bleat like that of the grey pup, but the adults have a sharper, less musical yelp or bark.

During the early years of their lives dolphins and seals are at their best for a harmonious relationship. They are companionable and playful in this period, and it is possible to explore their intelligence and cultivate a friendship by kind treatment and patient experiment. Usually the giving of food rewards is necessary to stimulate special responses. But neither Diana nor Opo required this in return for companionship; Opo refused all offers of fish or other food – she wanted only kindness and company and games.

All the evidence, sadly, shows that, like man, as they get older these sea-mammals of keen intelligence get less and less playful. In captivity they become more reserved and dignified. When they are really old they require, like us, only to eat and sleep.

We do not know how long they live in a wild, free state. Probably the majority die young—like those whom the gods love.

2 Calf Seal

Few people have witnessed the birth of a seal. The first parturition of a grey seal which we witnessed, from the height of a Welsh cliff, was completed in a few minutes. The mother's labour appeared not to be strenuous, as she reclined on the pebble beach near the entrance to a cave. An hour before birth she glanced frequently towards her tail end, obviously in anticipation. Then her body began to contract spasmodically, between peaceful intervals. The baby's nose appeared first, veiled by a membrane. Within ten minutes it was all over.

The calf was suddenly expelled in a final spasm which burst the membrane. The mother turned around swiftly, and in doing so broke the umbilical cord. The baby gasped the first air into its lungs, and at the same time we heard a feeble wail. The cow responded by nosing or smelling at her child, and touching it with one flipper. She made no attempt to clean or lick it, as many other mammals do. Nor do cow seals touch the placenta which usually slips out easily within an hour of birth. (Strangely, in herbivorous animals the parturient cow, ewe, doe, etc., will eat the afterbirth, which is said to help fill the void left by the exit of the foetus.) The cow seal ignores her placenta, which is devoured by scavenging sea-birds or washed away by the tide, or both. It contains the urine and faeces of the pup, as well as moulted hair, for the pup is already actively drinking and defaecating before birth.

Her retractile breasts, placed low down on her trunk, were already protruded and dripping milk. She rolled on her side so that the teats were quickly accessible to the calf, which was nosing instinctively in search of them within a few minutes of struggling free of the membrane. It looked as if the calf sought the milk-wet nipple by scent as well as touch.

Such we found was the usual behaviour of mother and pup at parturition. We have witnessed five births of grey seals. The very few observers who have seen a common seal born report a similar sequence of movements, which are evidently common to the phocid seals. The mammae of seals are retractable; at the touch of the calf's nose the nipples move outward, if not already distended in anticipation of suckling; they invaginate slowly after the pup has suckled, and this mechanism protects the mammae from abrasion during the normal land and sea activities of the mother. The pup's tongue is notched at the tip, probably to obtain a firmer grip of the small teat.

Judging by the apathy of the older cows, parturition is easy to the experienced mother of more than one or two pupping seasons, and painless. But the cow seal giving birth for the first time may experience difficulty, cry out, even become hysterical, as studies of the elephant and fur seals have revealed. A young grey-seal which gave birth to her first calf close to where I was encamped on a nursery beach 'moaned at intervals, ending with a louder, more prolonged wail'. This was on the Welsh coast but as it was midnight I could not observe her visually. But she appeared to be in severe pain during parturition.

The expectant mother is usually restless and irritable. In those nurseries where there are many other seals lying about, she will seek out a space or quiet spot in which to give birth. This restlessness is exhibited in her tossing around often, in snapping and moaning at trespassing neighbours. The grey seal will at this moment be accouched above normal high water mark, or even some distance inland. But the common seal bears its young one close to the sea, usually below high water mark. Both species will give birth on the ice, where ice conditions prevail in the pupping season.

It is believed that the common seal pup can, like the whales and dolphins, be born in the sea, and in that case it floats to the surface, where, like the dolphin baby, it is kept with its nostrils above water by its mother, or it keeps itself in this position instinctively, until it has drawn sufficient air into its lungs and venous system to oxygenate its blood. It does not drown, but floats buoyantly. We cannot find any directly observed account of this interesting event, but L. S. V. and U. Venables consider that in a rocky habitat in rough weather when it is difficult for a parturient common seal to come ashore owing to the heavy surf, delivery 'must often occur in the water. On June 18th, while watching two females with new-born young in shallow water, we saw a fresh pink afterbirth bob up to the surface close by. Four fulmars (*Fulmarus glacialis*) pounced upon it immediately and tore it to shreds. There was a heavy onshore swell at the time and it was almost full tide, nor was there any convenient beach nearby. Probably, under these conditions, the pups can be born at sea in these shallow sheltered bays.'

This observation is suggestive, but not absolute proof of birth at sea. The afterbirth could well have been expelled at sea by a cow some time after the pup had been born on land and had entered the sea when the rising tide washed it from the shore. Fisher found no evidence that common seals in the Ecstall river of British Columbia were born in the water; afterbirths were only found on the hauling-out bars which were completely covered at high water.

Does this evidence mean that the cow can control the moment of parturition until conditions of tide are favourable? Some writers believe that this is so, that a gravid cow can delay the birth of her calf for even longer periods. Oliver Hook (see page 151) considers that the grey seals breeding in the Baltic in late winter may 'delay parturition pending access to suitable breeding ice'. On the other hand it is recorded that, in large colonies of elephant seals (a related phocid), some cows are overtaken by parturition on their way between the sea and the inland nursery harems, in which case the pup so born is usually although not invariably abandoned – the maternal instinct at the moment of parturition is weaker than the social drive – and the pup is left behind, moaning and starving, unless birth occurs within a few yards and in sight of the harem group.

J. L. Davies states that in grey seals on Ramsey Island, Pembrokeshire, most births took place when the tide was rising, a convenient time for the gravid cow to haul ashore, since at low water much of the exposed tidal area consists of rough boulders difficult for her to negotiate; but dangerous if the tide is high springs and the sea stormy. In the five births I have witnessed the tide was half-ebb (3), half-flood (1), and low water (1) respectively. These all took place close to or above normal high water mark, the cow having in each case been accouched there for some hours beforehand.

Although grey, common, elephant and fur seals have been seen suckling two or more pups, proved twins are very rare, but occasionally recorded from the presence of two foetuses in females killed during pregnancy. Triplets are unknown. The late Frances Pitt wrote to me after a visit to Ramsey Island in October 1948: 'As I was scrambling down to the little beach where there were four very young calves lying under the right hand rocks, a cow grey seal came ashore and scrambled towards them. Two of the youngsters, identical in size and appearance, met her, she nuzzled them both, then rolled over on her side and both sucked – at least both had their noses firmly against her. I had a ciné camera with me and got a good record of the episode before she saw me and bolted back to the sea.'

E. A. Smith has informed me that in the crowded colonies of grey seals in the Orkneys he has frequently seen two calves sucking the same cow, both together and one after the other – this was in a large congested rookery. He considers that a cow with plenty of milk may tolerate the sucking of one or more alien calves which are hungry enough to approach her; even that calves which have mislaid their mothers may suck a foster-mother by stealth (as domestic calves will), perhaps to be driven off by the cow when she discovers the culprit. He also reports finding – a rare instance – two embryos (one dead and one dying) in the same set of membranes of a pregnant cow. The incidence of twins in the Alaska fur seal is only 14 per 10,000 gravid females, or 0.14 per cent.

The sex ratios at birth in all seals appears to be the same: 1 : 1. But in grey seals there is a slight preponderance of males born early in the season, with an equalising slight preponderance of females later.

The majority of pups are born head first, but in a number of cases breech or tail may be presented first – the hind flippers appear and the foetal membranes which are pressed against them in labour may be prematurely broken by the long sharp-pointed hind claws.

The genital organs of the new-born seal are, like those in man, in a state of physiological stimulation from the endocrine substances contributed by the mother through the placenta. The ovaries are enlarged, exceeding in weight those of the mother, and the uterine glands resemble those of a bitch on heat. In male new-born pups the gonads are large and heavy. But this gross condition, resembling that of normal oestrus in the adult, is temporary. There is a rapid transformation in the first ten days, at the end of which the ovaries have shrunk to one-eighth in weight and the male gonads become tiny and dormant.

The cow recognises her calf by the use of three senses: sight, to locate it at a distance; hearing, by responding to its infant cries; and smell, which is the final test when the two touch each other. A cow will be seen working her way through a group of calves, smelling and pushing aside or ignoring each in turn until she has found her own child by scent recognition.

The resident bull in both species ignores or shows no interest whatsoever in the calf. The calf is likely to snarl at a trespassing bull, which may retreat as a result. But if in search of an oestrous cow, or intent on attacking a challenging bull, a resident bull will ride over the top of a calf as if it was not there.

The grey seal calf at birth weighs between 27 and 32 lb, with extremes of 20 and 40 lb. The new-born common seal weighs about 10 lb less; on average 22 lb, with extremes of 15 and 39 lb. Contrast this with the weight of the elephant seal at birth, about 83 lb; female pup elephant seals double this birthweight in eleven days, treble it in seventeen, and quadruple it in twenty-one days. Male elephant seal pups are usually several pounds heavier than females at birth, and quadruple their birthweight in the same time, about three weeks, which is the approximate date of weaning.

Grey and common seal pups do not grow at quite this rate, but normally they double and sometimes treble their birthweight at weaning.

Common Seal Calf

Common seals the world over are born in the season of the best weather of the year for survival, around midsummer. The pupping season begins early in June, and lasts about three weeks, with occasional early births in May, and late ones in July. On the Wash sandbanks the fishermen declare that there is a general 'fall' (birth) of pups at low tide on midsummer night (June 21st–22nd) which is approximately true.

The new-born pup is lean, lanky and long – usually under three feet, between 700 and 800 millimetres, from tip of nose to tip of tail. It is sometimes covered with loose yellowish-white foetal hair, which disappears within twenty-four hours. Most of the common seals in British waters have already moulted before birth; the white foetal hair is ejected with the placental bag, mixed with prenatal urine and faeces.

In the Wash on June 25th I examined two new-born brown-coated pups on a soft sandbank, with placenta alongside, helpless to follow their mothers a few yards into the sea, although, as our small boat approached, their dams made frantic rushes in and out of the water, trying to coax the uncomprehending infants to the safety of the sea. The agitated mothers made no attempt to drag the pups to the water (although this has been reported by other observers occasionally) but swam at a distance as we grounded the dinghy and examined the pups. One was still attached by the umbilical cord to the placenta, but appeared to be breathing normally. It did not move away, but looked up at me, sniffing the air with raised head. The other shuffled a few inches towards me, and made no resistance when I rolled it over to find the umbilical cord severed at about six inches from the belly. As I moved away this pup struggled feebly to follow me, bleating pathetically.

Some thirty adults and half a dozen lively pups had escaped to the sea on our approach in the dinghy, leaving behind one dead pup. When we returned to the boat and motored out to sea the anxious mothers of the two live pups could be observed through binoculars hastening on to the sand beside their babies before any other seals took the ground.

If undisturbed by man or tide the pup will lie close to its dam for a few days where it was born, sucking milk at frequent intervals, and rapidly filling out its sack-like body. It can be captured at this stage easily, since it finds great difficulty in moving about on land. The front flippers are very short, and the hind-flippers no use at all. Nor does it recognise man as an enemy at first, but will remain ashore when danger threatens, as the pups quoted above did, sleepily ignoring the attempts of the cow to move it to sea on the approach of a boat. At this age the innocent pup will quickly adopt a visiting human as its parent and try to follow him or her into the sea. In Norfolk, visitors who find healthy seal pups ashore near their breeding grounds are warned by the RSPCA officials not to handle or fondle them, lest they transfer their dependence on their dams to humans, who are usually quite unskilled, unable, not to say unwilling, to rear them. The RSPCA has lately maintained a summer crèche on the shore in East Anglia where lost pups are cared for, fed by experts, and later released when considered able to fish for themselves.

The majority of pups, living with their mothers, however, learn from the action of the adults in taking them to sea on the approach of man, that he is a potential enemy. The adults will nevertheless become accustomed to fishing boats using the area of the sandbanks regularly, and will haul out near the boats provided the occupants do not molest them, or stand around in conspicuous attitudes, holding guns. Fishermen have many times told me that a seal knows when a gun is in the boat, and will not tolerate the presence of a boat from which they have been hunted previously. They vanish before such an armed boat can be brought within effective killing range. I am not surprised by this assertion, for I have found the intelligence of the seal as high as its capacity to learn from experience.

The new-born pup progresses very slowly on land. It squirms along by raising its head and upperparts in the air on the levers of the fore-flippers which are pressed into the ground with the webbed digits flat, or else with the nails gripping the ground. The body then topples forward, the breast often striking the ground, and the fore-flippers are 'rowed' forward for another heave. The tail end is simply dragged along limply, or with a slight twisting movement as in swimming. (See Figure 2, page 74.)

Once in the sea it swims gracefully and dives with ease. But as yet the sea can be a principal danger for the very young pup. Severe ground-swell may sweep it against cliffs, and a still more frequent cause of mortality is separation from its dam by strong-running tides. Every year the strong tides of the Wash area sweep some of the younger pups away, and those which are unaccompanied by their mothers are lost and eventually die of exhaustion and hunger if they are not, as sometimes happens, picked up and cared for by human rescuers. These lost waifs become very tame with hunger, and will 'come sucking around the boat', as one shrimp fisherman graphically described their attempts to find nourishment from the curving belly of his little boat. They will also try to suck each other, even taking the stub of the umbilical cord of another pup in the mouth.

But generally the mother does not easily become separated from her child. When hungry, and without its mother, it utters a fairly penetrating wail, and this will guide the cow back to it. When man approaches by sea or land she will coax it into the water. She will jealously cradle it in her arms as it takes its first swimming exercises, perhaps manœuvre it on to her back to relieve its tiredness and keep its head above water. Mother and child are delightful to watch. The pup has three main urges: to feed, to sleep and to play. Much of the time in the water is spent in playing with its mother – good training for the young muscles. She will dive with it, at first cradled in her arms, later on her back. When sleepy the pair will float at the surface, side by side, heads slightly below the surface, above which only the hump of the back protrudes; in this attitude they look like dead, drowned objects.

In the Shetlands, in a rocky habitat, the Venables observed that the pup had difficulty in getting out of the water on to the tidal rock on which its mother was resting, owing to the weak traction power of its limbs already mentioned. The wave which washes it up on the rock washes it off again. But a pup may be able to suck milk from a cow resting on a low rock by reaching the teats while its tail end is still afloat in the sea, on a calm day.

The Venables declare that until it is about three weeks old the common seal pup cannot hold its head above water for long periods;

and that when resting at sea it does so with only its crown and shoulders above the surface; the nostrils are lifted to breathe air at regular intervals; and in this curious 'hump-back embryonic' inverted-U attitude it is perfectly at rest in the shelter of a calm bay. But grey seal pups also do this, and so do adults of both species. Pups we have reared would rest in this attitude in bath or pond when sleepy.

The permanent teeth are already through the gums at birth, and grow rapidly during the lactation period, presumably to be ready for use when this short phase is over and it is quickly necessary for the pup to hunt for and catch solid food. The little teeth do not incommode the mother, for the short teat is taken between the notched tongue and the gum between the canines in the upper jaw; yet the grip is firm, so firm indeed that the baby is sometimes towed along in the water as it holds to the teat. Suckling is carried out wherever the mother and child meet – on land, in surf, but most often in the water. The dam may float vertically at the surface, the pup is usually head down under the surface as it sucks, its tail often the only part above water. When the mother floats horizontally the pup is upside down, attached to the teat with its head beneath her.

The stub of the umbilical cord, from two to eight inches long, remains fresh for two or three days if the pup is in the water, but withers quickly and drops off if the pup remains on land during this period.

The mother's devotion to her child continues for the first two weeks and although it lasts to the end of the lactation period of four or five weeks it slackens during the latter half, when she guards her pup less tenaciously. It has become fat and round-bellied, and begins to haul out alone and sleep for longer periods ashore. Its mother plays with it less often. Instead the pup plays with other floating objects such as a plank, piece of driftwood, a fishing float, or seaweed. Tame common seal pups, such as Diana, loved to play with any movable thing they could find in the water. They invite their keeper to play with them in the water; they love to be rolled over, rocked, stroked and talked to. They are intensely curious; and may get into trouble by swimming up to a boat containing hostile fishermen. 'On such occasions,' writes Fisher, 'the mother seal would frequently surface suddenly beside it, and in a flash grasp it by the back of the neck with her teeth and pull it under water in the middle of its typical cry, in which case a distinct gargling sound ensued. Lost pups keep up a steady bleat until rescued by their dams. The gargling sound indicates that the mother has rescued a pup in danger and dragged it away to safety under water.'

Unlike the solitary unsocial grey seal pup, the common seal pup soon graduates from playing with its dam, and with driftwood and seaweed, to playing with other pups. There are congenial games in the kindergarten of the weedy shallows of the summer sea, and afterwards the suckling pups will lie ashore in little groups, stranded well above low water mark. Experienced seal hunters wait for low tide to visit the pupping grounds in late July and August. They can then easily kill, by clubbing or shooting, these four-week-old pups, by intercepting them as they crawl towards the water in alarm. They are slow to escape because of their failure to perceive man as an enemy until too late, and because of their immobility ashore. Shetland hunters also shoot them at short range as they float, hump-backed in the water, asleep and innocently unaware of danger. At this age of a month the coat is probably at its best.

Soon after, perhaps during the fifth week, the cow's milk has dried up and she is no longer interested in her pup. This is about the duration of the lactation period in the Wash and in the Shetlands. In the Fraser river of British Columbia Fisher found milk in the stomachs of pups in August, and even in three stomachs obtained on September 1st, which is above eight weeks after the normal fall of pups: he concludes that there must be variation in the length of the lactation period, probably due to feeding (nutritional) differences.

We have seen, then, that the common seal pup is quickly adjusted to a wholly marine life. Our captive individuals could sleep at the bottom of the pond within three weeks of birth, resting with eyes tight shut for up to three minutes before rising, still apparently asleep, to breathe for about half a minute before submerging again. It is not difficult to imagine that, like our captive individuals in a pond, a wild-born pup, newly weaned, will begin to chase and eat what it can catch of small fry below the surface, when sufficiently hungry. Prawns and shrimps and small rockfish are recorded from the stomachs of weaned common seals taken in both European and North American waters.

The weaned pups do not wander far from the area of sea, river estuary and sandbanks frequented by the sedentary adults. Although independent they associate with the grown seals, which are now, in late summer, engaged in their annual moult.

The pups do not moult at this time. They have already moulted the white coat before birth, and are clad in a dark one. This now consists of a mixture of thick, rather coarse overhairs about one centimetre long, covering a dense inner stratum of hairs almost as coarse but only half as long. Each overhair has four or more underhairs attached to its root. Such an arrangement gives the young seal immediate protection when it enters the water; and in addition it has sweat-glands which exude an oily fluid, assisting in the waterproofing of the body. The common seal is fully insulated against cold, and is one of the most aquatic of seals at birth.

There is great variation of coat colour and markings, and this variety persists through life. 'Almost endless,' remarked Allen in 1880, 'varying from almost uniform dark brown and even nearly black with, between these extremes, almost every possible variation, from dark spotting on a light ground to light spotting on a dark ground. The markings vary in size from very small spots to large irregular patches and streaks.' Males generally are darker and more spotted than females on the flanks and belly, and about the cheeks and throat. But this is not at all invariable – one can find a female heavily spotted all over. The claws of Dick, an orphan male we reared, were black, and so were his whiskers or vibrissae; but those of Diana and other female pups were as pale as the human nail is, and the female moustaches were whitish.

At a month old the pelage of the common seal is very handsome. When wet and smooth it looks sleek and almost black; when dried out it is browner because the underfur then shows. The young coat is thick, close and more uniform and finer to the touch than the coarser worn hair of the adult. At this age it is in demand for sale to the furrier; when cured it is soft, durable and free from blemishes.

Grey Seal Calf

The grey seal calf lies almost helpless for a while after it is born. Its head, heavy in relation to the weight of the slim sack-like body, rests prone on the ground. It is lifted with difficulty. The hind flippers lie flat on the ground, side by side, as they will never be again in life; they gradually become vertical as the body fills out, barrel-wise, during the lactation period. If disturbed the new-born calf can drag itself downhill, but it cannot lift more than its head and chest clear of the ground, and at this tender age of a few hours it cannot climb up even a moderate incline. If by any chance it is swept into the sea at birth it will drown.

Usually however it does not have to engage in violent exercise. The mother delivers the pup above high water, and under ideal, undisturbed conditions remains with it for two or three weeks, feeding it every hour or so at first, whenever it is hungry and seeks the breast.

At birth the grey pup is covered with long white hair (about thirty millimetres in length) which is firmly attached for a day or two, and this white pelage is prized by the hunter as a marketable product. But the pup must be killed at or soon after birth, for the white hair quickly loosens and begins to fall out, and occasionally the moult begins before birth. The white natal fur is stained with the cream-yellow amniotic fluid of the placental bag, and remains discoloured unless rain or sea-water washes it white.

Occasionally a grey seal mother will seem remarkably indifferent to the new-born child. In Wales I have watched a cow leave her pup immediately it was born and with no more than a casual backward glance go down to the sea without giving it suck. She appeared to be nervous, although she could not see me watching from a niche in the cliffs above; she may however have been fearful of past intrusions by visitors on this particular beach. Her pup moaned and struggled inch by inch down the pebbles, but halted at the edge of the sea. Next day I found the mother back with the pup, which was sucking contentedly, and was eventually reared. After suckling the cow patted and scratched her child gently with her fore-flipper, a pleasing sign of love which I have often witnessed.

The wail of the hungry pup is pathetic – almost like that of the human baby, but difficult to render in words. Rather more musical than that of the common seal pup, it resembles a long drawn *Sōh-rāāb*, uttered as the pup's mouth is opened wide, the first syllable as the jaws fall apart, and the harsher *rāāb* at their widest. The cow rarely calls in reply, but, guided by the wailing of the pup, she makes her way over the beach silently to her child.

The wailing of hungry pups which have mislaid or wandered away from or been deserted by their dams is characteristic of a large seal nursery of several hundred pups. Such assemblies or 'rookeries' occur on certain small islands in the Orkneys and Hebrides, where the breeding bulls and cows haul out over grassland for many, occasionally hundreds, of yards inland. Here, with each cow usually attached to the harem of a dominant bull, the calf is safe from the sea, but faces other hazards. It may very easily stray from the cow during the boisterous territorial activities of the bull, and fail to return. Master bulls, in fights with challenging bulls, or in love-making affairs with the cows of their harem, may blunder over the top of a new-born calf too tender to withstand such bruising, and before it has acquired a protective cushion of blubber. The pup may be thus accidentally crushed and die. Or, alarmed at the rough behaviour of the adults, it may slither away, perhaps to fall into a fold of the ground, a stream, or trench used by the adults in traffic to and from the sea. If it does not manage to get back to its mother the pup will die of starvation, or of being crushed in the traffic lane. Mortality from these and other causes at a rookery is between 10 and 20 per cent of pups born.

Normally the cow guards its pup constantly, and will drive away other cows and bulls. Some cows protect their pup with quite ferocious solicitude, and I have been driven off a nursery by a bold cow more than once. To me, having watched both species with their pups, the grey cow's love is rough and aggressive, compared with the tender and intimate love shown by the common seal for its child. And from its mother the grey calf learns that aggressive behaviour with which it repulses the advances of neighbour seals; we have seen this behaviour in the relations between the friendly Diana and the unsociable Sabrina.

The grey pup retains its white coat for about three weeks, frequently less, often longer. The majority are in full, or almost full, grey coat at four weeks of age, by which time they are quite deserted by the cow, whose lactation ends in the third week, earlier than that of the common seal. The pup is now known as a 'moulter', although in fact it has already moulted.

At birth the grey seal pup weighs round about 30 lb. It is fed on the rich seal milk, which contains approximately 50 per cent fat, which is ten times richer than Jersey cow milk. At the end of the lactation period it is extremely fat and weighs between 80 and 110 lb, indicating an average gain of 3 lb a day. The cow in this period remains fasting and loses nearly double that amount of weight daily, that is, she loses about 100 lb in about three weeks, a formidable drain on her resources, and only possible because the lactation period is short, and she can begin feeding at once after deserting her pup. In fact her lactation period is as short as that of a mouse – twenty-one days!

We have already described how grey seals, born close to the sea in caves and open beaches, learn to swim within a day or so of birth, but they do so unwillingly and only when they are forced by a high tide flooding the nursery, or by disturbance from man. If so compelled to enter the sea, the pup is normally attended by its mother. She will endeavour to coax it ashore again, but if storm waves or tidal currents carry it to sea she swims with it. She is rather helpless to guide it, less apt to clasp it in her flippers than the common seal; nor does the pup make much attempt to hold on to its dam. At this tender age it does not seem able to grip well the surface of its mother's smooth body, although later it will ride pick-a-back.

When storm-waves on a rising tide crash on a narrow nursery beach the grey cow will often lie full length along the seaward side of her pup, breaking the force of the surf which might bruise her child. I have seen her drag it out of danger by seizing it by the ankles of the hind flippers and flinging it backwards up the beach and behind her, as storm waves broke on the pup, which had crawled to the sea when the tide was low. Her sharp teeth cut the ankles quite deeply; many pups bear the marks of the cow's teeth on this part of their body, at least in Pembrokeshire. On two occasions we have seen her seize her pup by the scruff of the neck and move it up the beach. But perhaps she is less expert than the common seal cow at handling her baby at moments of crisis. Only rarely will she cradle it in her arms at sea, but we have seen this happen; on one occasion during a heavy gale when she was trying to keep her pup away from a lee shore on which already several pups had been washed in, drowning or drowned. Behind this beach was a hamlet with dogs, and the cow seemed to be aware that the beach was not safe from human interference, although otherwise offering a temporary resting place on sand as the tide ebbed. On other occasions I have seen the cow cradling the pup threatened in a swell against a rock-bound coast over high tide, and an onshore gale blowing.

Like the common seal, the grey seal pup will suck milk above, on the surface or under the water, although normally suckling takes place on dry land.

For a short period – the latter half of the lactation – the beach-born grey pup enjoys playing much the same games as the common pup plays with its mother in the water, on calm days, an experience which the pup born far inland will never enjoy: follow-the-leader, pick-a-back, and hide-and-seek.

But it is a very short period of a few days. The cow has mated at the end of the second week, and as her milk dries up she loses interest in the pup, and presently, a few days later, abandons it. It is an association never to be renewed, a final parting. The cow leaves the nursery area; the calf seeks out a quiet place in which to rest and complete its first moult above high water mark. Here it remains, if undisturbed by man, to complete the last vestiges of moult in or near its birth place. At this age of four weeks the dappled light or dark grey coat is as handsome as that of the weaned common seal pup, and generally more boldly marked. But the variation again is tremendous. The darker spots and blotches are larger and more distinct in the grey pup, and the bull is usually darker than the cow – as in the common seal. Some individuals have cream or buff bellies, and the grey pup at this stage is less brown beneath than the common pup.

During this stage of fasting the grey pup loses approximately half the weight – about 70 lb over the birthweight of 30 lb – gained during the lactation period, but continues to make bone and length. It is approximately 65 lb in weight when it takes to the sea.

The 'moulter' pelt of the grey pup, free of moult now, with its short, close, beautifully patterned new fur, is prized by seal hunters whenever they can find and take them. This is fairly easy if weather permits a landing in October, November and December on remote islets where the pup, retreating well inland during this period away from the turmoil of the breeding nursery, is at the mercy of the hunter. It is sluggish and often makes no attempt to escape the hunter cautiously approaching from seawards; but instead looks sleepily up at him, and awaits the unexpected death blow or shot with wide, huge, innocent, brown eyes. Should it become uneasy and make off towards the sea, it is easily intercepted and slaughtered.

However, where it is born on a beach at the edge of the sea, it will early have learned to swim, within a few days of birth; and by the time it is independent at three weeks of age it may have acquired to the full the normal dependence of the adult upon the sea as a refuge from man. The hunter who would take a moulter lying close to the sea knows well that he must stalk and surprise it before it bolts into the water.

3 The First Year

The study of mammals, including man, reveals a general pattern of behaviour common to each principal age-group – infant, adolescent, adult. After weaning and before puberty there is a period of restlessness and wandering. Young people and other young mammals have an innate desire to leave home and explore the wide world. They may depart voluntarily, or they may be driven forth by the adults which monopolise the home territory and its food supply for themselves or for succeeding broods. There are sound reasons for this dispersal; in particular, economic factors are at work. In man there is the need to earn a living; in other mammals there is the same need to search for food and shelter away from the exploited home sources.

In the grey seals marking of individuals has brought returns which confirm this annual outward explosion of population. It is the young ones under a year old which wander farthest from home – although it must be noted that such marking is biased in favour of recoveries of these youngsters because few mature seals have been marked, and because marking tags, rings and brands tend to become lost or unreadable after a few years; also mortality, leading to recovery of the body of a marked seal, is naturally highest in the youngest age-group.

The period of restless wandering terminates, or else assumes a regular geographical pattern in those mammals which perform annual migrations, when the individual establishes itself in its first breeding ground, either at or just before mating, or at or just before parturition.

Comparatively few common seals have been marked, but it is

generally assumed (from reports of fishermen, for example,) that the species in all its age groups is sedentary, that the pups in their first winter do not wander far from the birth place. There are feeding movements of the herd up and down estuaries, and along coasts and familiar sand bars. But individuals of all ages from weanlings to adults may be seen month after month hauled out together on certain favourite banks and rocks. The Venables studies describe the common seal herd in the Shetlands as static and faithful to one place throughout the year. This sedentary habit may explain why *Phoca vitulina* has become subspeciated (that is to say, has been divided arbitrarily by man into five subspecies) in its breeding grounds across the waters of the northern hemisphere, whereas the more wandering *Halichoerus grypus* has not.

Nevertheless, its wide distribution as a species, covering the northern Pacific and Atlantic, the southern Arctic, and certain fresh-water lakes, seems to indicate that the common seal has not always been so sedentary that individuals have never wandered from the home waters and set up new colonies – which, adapting to new conditions, have become (on paper) subspeciated. We know that from time to time young common seals appear in localities out of their normal breeding range. One was seen at Skokholm Island, Wales, in September 1963, some eighty miles from its nearest known group. We have mentioned the tame individual which frequented Cuckmere Haven, Sussex, although common seals are not normally found in the English Channel. They once bred as far south as the Somme, up to the end of the nineteenth century, according to Lucas: he reports a very young common seal, 700 millimetres in length, taken alive at Cancale in September 1955, which died soon after capture – obviously a wandering individual only two or three months old. Individuals have also been reported, though rarely, from Spanish waters. At one time the common seal bred in the Faeroe Islands, but is now a rare visitor, apparently extinct as a breeder along that wild inhospitable shore where there are no broad sandy estuaries to its liking – but it inhabits Iceland. Full distribution details are given on page 153.

The weaned grey seal pup, by comparison and on the evidence of a dozen years of marking in Great Britain, is at first a wanderer. Although in our experience in Wales some weanlings may be seen around the nursery coast all through the winter, marking has shown that in its first six months of life some individuals travel up to 600 or 700 miles from the birth place. One Pembrokeshire-born pup reached Santander, northern Spain, at the age of ten weeks, indicating that it had travelled about 600 miles during the four or five weeks between leaving the nursery and arriving, alive and well, in Spain, a remarkable swim for one of such tender age – allowing for the normal deviation and feeding activities while moving at sea, it had probably swum a total of thrice that distance before reaching Spanish waters.

This record is paralleled by the swim of 600 miles made by a grey pup, marked on the Farne Islands on November 5th, 1959, which was recovered (dead) near Trondhjem, Norway, eight weeks later. Another Farne pup has reached the Faeroe Islands. Grey pups marked in the north Scotland colonies have swum to Norway and the Faeroes, even to Iceland. Welsh pups have been reported in a wide dispersal ring, from the western coasts of Ireland, the Irish Sea, Brittany and the English Channel; about 8 per cent of all seals marked in Wales have been recovered, dead or alive, half this number over fifty miles from birth place (see map on page 159).

We have said that moulting of the white natal coat of the grey pup is complete by the end of the third week. Some precocious pups moult earlier than this, and others may take longer. Those pups which have taken early to the sea, through force of geographical, tidal and other disturbance factors, lose the white hair early, in swimming and intertidal activities. But in those born and suckled well above high water mark, undisturbed by man and remaining where they were born without going to the sea, the white hair drops and is scratched away, forming a downy ring or mat of fur around the plump body. Such pups are usually in fine heavy condition when they are abandoned by the cow, who for two to three weeks has acted, so to speak, as a reservoir of highly nutritive milk ever on tap. The pup has not wasted its energy in battling with tides and rough seas. It has sucked, slept, stretched, scratched and lived in perfect security in one spot a yard or so square. And it has remained there after the cow's departure for yet another week or ten days, fasting and converting the thick coat of blubber into bone and muscle.

When the grey-coated moulter which has never been to sea is examined for the first time it will gaze at man with what one may imagine is a puzzled look. I have stood still beside a fat sleeping moulter and studied its reaction to the upright human form on waking. Provided I remained perfectly still its response was mild and full of curiosity. It leaned towards me, neck stretching, nose and whiskers twitching to get my scent. Gentle slow movement of my arm aroused greater interest, but no alarm. Sudden movement, or an attempt to caress or touch it brought the familiar response of snapping and hissing with open mouth, the usual reaction of the older grey pup, which through the example of its mother has learned to distrust every other large animal, including any other seal.

Any persistent attempt to handle the moulter, as for purposes of marking and measuring it, will result in fierce resistance, biting, and usually the young seal will endeavour to escape seawards. It is heavy, powerful and difficult to hold – the seal marker needs his whole weight to do so. The seal marker must usually sit astride its back, pressing its head to the ground to avoid being bitten, while a second operator attaches the ring or mark.

The grey pup remains unsociable and hostile to all other seals on first becoming independent, in sharp contrast with the friendly, playful common seal pup. It is however gregarious to the extent that it will haul out in the company of other seals, but keeping a wary distance of a yard or so from its nearest neighbour. It has, as already suggested, acquired from its dam this habit of preserving a 'safety distance' around it; she established such a safety distance or 'pup defence area' around herself and child from the beginning, attacking any other seal, of any age or sex, which tried to invade that magic circle.

At the hauling-out rocks or beach the moulter gradually discovers its place in the social hierarchy or order of precedence ('peck or bite order'): on the bottom-most step. Depending on its strength of character and physique it may or may not succeed in driving away a seal of its own age. Usually the seal already in possession of a desired lying-out site will succeed in retaining it against an invading seal of the same age; possession is nine-tenths of the law – at least it seems to confer a psychological advantage, as well as the obvious moral right. The newcomer will veer away at the moans, snarls, and flipper-waving with which a seal warns an intruder to keep clear.

Grey seal adults will ignore these youthful warning signals if they desire to reach a comfortable spot on a beach, and will plough towards it, scattering the pups and yearlings with snarling noises, uttered from wide jaws; and by this threat behaviour will obtain the basking territory without actually biting a pup or a yearling. The youngster may shuffle off to the sea, but usually it moves aside to the nearest vacant spot, and may be tolerated by the usurper at about one body's length away.

This is a tough period for the young moulter, of adjusting to a life of fishing at sea, and capturing solid food in place of being fed on breast milk. Many do not survive. It is estimated that 60 per cent of the young phocid seals die, from all causes, in the first year. This figure is given for the Alaskan fur seal after some fifty years of study and commercial management of the Pribilof herd, during which approximate longevity and other vital statistics have been worked out. Mortality in the grey seal, believed to have a similar life-span, must be close to this figure. About 20 per cent of pups die, or are born dead, at the nursery (Farne Islands, North Rona, Orkney, Pembrokeshire) within three weeks of birth; leaving 40 per cent to perish before they are a year old, from dangers at sea – weather, hunger, and predators.

Very little is known about how the young seal begins to feed and fend for itself, but from an examination of the stomachs of under-six-month-old individuals of both species, it would seem that in their first independent swimming and diving play they find small edible fish and molluscs, including shrimps and crustaceans, prawns, crabs, octopus, squid, and rockfish. Our tame Diana and Sabrina and other captive pups were uninterested in dead or unmoving fish, but immediately gave chase to any live thing that moved under water; and, as we have already said in the last chapter, one can well believe that the young seal, when hungry enough, will catch and eat the small fry of the shallows as it darts before their bold eager eyes. By trial and error they no doubt graduate to those larger fish which they find most satisfying to eat. In captivity seals show a definite preference for certain fish: grey seals like flatfish, octopus and mackerel; both species eat herrings with great appetite.

After its wanderings, perhaps for hundreds of miles in coastal waters, the grey seal pup at six months old, greatly reduced in numbers, can be seen best in the following spring near the breeding grounds. It is evident that a proportion return at this age, and haul out in the mixed assemblies of all ages which are at their height from late February to mid-April (Wales).

These six-month-old grey pups have undergone a change of colour although not of coat. They are very distinct: the fur, even when wet, is pale, ochre-coloured, dun or light brown; the spots and blotches so faded that they do not show up at a distance, although the pigment is apparently permanent in the skin beneath (a seal cannot change its spots any more than a leopard); and when the fur is quite dry at the haul-out it appears paler still, almost cream-coloured and conspicuous accordingly. Like common seals at this age they are lean, fast-swimming and agile, and weigh little more than they did at three weeks of age, about 100 lb.

They are still shy of the older seals and to some extent of each other. Generally they keep apart, alone, in the hauled-out herd. But occasionally they will play together in the sea, rarely or never on the land. Over the second six months of their lives, covering their first spring and summer, a time of fine weather and increasing abundance of fish food, the young seals lose much of their acquired fear of and hostility towards each other, and learn to play together more freely in the water below an assembly beach.

After its initial period of wandering during the first winter of its life the grey seal becomes more or less sedentary. Already it is attaching itself to a locality or stretch of coast where it will remain for the rest of its life without making more than local feeding and moulting and breeding movements. There is no evidence from the ringing recoveries of further wandering into distant waters. Neither the grey nor the common seals are pelagic, like most of the whales and dolphins, and some seals (harp, fur, elephant seals). But as a result of the first winter wanderings there is evidently a redistribution of population, with fresh blood reaching some colonies in the form of young potential breeding stock.

The adult grey seal moults early in the spring, the adult common seal late in the summer. But the grey pup, born in a white coat, and moulting this within the first month, retains its first grey coat for more than a year, during which it fades as above described until it is quite dun and drab-coloured.

But with its second grey coat, acquired during the late autumn or early winter, the year-old grey seal is once more handsomely marked and richly coloured. The young male is darker in this second grey coat, and his spots and blotches have expanded in size with his expanding skin. Generally, the male grey seal can be described as having a dark coat with light spots; and the female a light coat with dark spots. Both are paler underneath, the cow having a creamy neck and breast. In faded moulting coat the colour, when the fur is dry, is drab brown, beige or tawny. When wet it assumes a much darker hue.

The common seal is also a year old when it begins to shed its coat soon after midsummer.

During the moult the young seals haul out a great deal on certain favourite reefs, skerries, beaches or sandbanks, where they are more or less undisturbed by man. These are traditional sites, common ground to all local seals, and known from time immemorial to man, but inaccessible to him except in calm weather, when, because of their open situation, his approach can be seen by the seals. Many of these sites are on islands, but some are in deep caves (grey seals) or under remote high cliffs, and known to only one or two watchers. Much of the observation carried out on younger seals and described here was made on the Red Wilderness beach, which sometimes holds over 300 grey seals of all ages; here I lived for a while, under great cliffs, while writing the book *Seals and the Curragh*, and have spent many hours there each year since I first discovered this hitherto unknown assembly in 1942.

4 Adolescence

Seen from the cliffs above the Red Wilderness assembly beach in the spring, the play of the eighteen-month-old grey seals is pretty to watch. It is much like that of the near sexless play of half-grown pup-dogs. They roll in each other's arms in the surf, biting playfully at each other's face and neck, climb over each other's back, and splash each other with fore-flippers. They enjoy their games enormously and for hours at a time, keeping to the surf as a rule, as this adds an element of chance to their mobility with its own movement of tumbling waves, forcibly initiating new attitudes and fresh positions in their wrestling matches.

On the same beach you may watch the play of the older adolescents, thirty months old, the pale-coloured virgins with the lanky black, new-moulted males; some of the latter may be a year older still – three and a half years. Their play is more powerful, intimate and persistent. The lean young moulted bull will straddle the virgin awkwardly, playfully necking her, mouthing her face, and endeavouring to curl his hind flippers around her tail. Now and then the bright pink male organ emerges from the penial opening in his belly, but instead of submitting quietly to him, the virgin, without sign of fear or attempt to escape, continues the ardent wrestling, endlessly trying to reverse position and mount the young bull. But in doing so she inadvertently, or perhaps deliberately, avoids copulation.

Desire and ardour – libido – are there, but weak, as if neither knew what forces were moving them in this spring-time game. In this pre-liminary embracing, this pre-play before consummation, the joyful courtship might continue for an hour, and end indecisively. At times the pair appear to be in coitus for a few seconds, rarely longer.

Often the embraces are continued from the beach into the sea, as one rolls and swims away, pursued by the other. Often it is the virgin who follows the bull, and initiates a water-dance, deliberately pushing her nose against his face in what I can only describe as a kiss. They embrace, arms touching, breast to breast, nose to nose. The virgin might fondle the bull and clasp his back, or if he lay – a frequent posture – lengthwise at the surface, she might attempt to ride upon his back. This might cause him to turn and bite her face, either playfully, or irritably if he was tired of her antics. Frequently they would sink together, clutching and mouthing, and disappear for a few seconds, or minutes.

Of 300 seals assembled on the Red Wilderness pebbles, and in the surf, there might at any given moment in February, March or April be as many as twenty couples playing this spring game in the surf or close inshore. The rest of the assembly would be sleeping, yawning and scratching, dry upon beach or basking rocks. There would be many mature bulls and cows in this herd, and a few of these might also be at play in the sea, swimming in pairs, rather more dignified than the youngsters, yet engaging in pleasurable manœuvres, endlessly gyrating about each other, caressing, and 'kissing'. In the dances of the adults also it was noticeable that the cow made the advances, and even sometimes two cows approached the same bull playfully. The adult bull is very much less ardent in the spring courtship. It is almost as if, while willing to humour the cow and play with her, he could not be bothered to achieve the position required for a successful coition while the cow fooled around him. This was more obvious when the adult couple rolled in the surf; the bull's penis might be unsheathed, but his struggles to position himself for mating would be frustrated by her skittish wriggling and attempts to mount him, and in addition the wave breaking over them might upset them at the critical moment. It is easier for the pair to mate in the water, which sustains their great bulk; and from time to time I recorded this in the spring from seeing the pair floating in coitus, the bull clasping the cow beneath him, and both motionless for long minutes, except for a lift of their heads to breathe from time to time.

As we shall see later in this book, even if the adult bull copulates successfully with the adult cow at this season of the grey seal's annual cycle, he may not inseminate her with fertile sperm. But for the moment we are concerned to describe behaviour in the spring assembly, in which all ages are gathered, and there is great amity in the herd. Except when competing for a basking site on beach or rock, there is no quarrelling even between adult males. In the water they swim apart, with at most no more than a threatening yawn at each other. Younger bulls actually play with each other freely, much as virgin and young bull do. Their homosexual behaviour resembles that of many mammals of adolescent age. In the grey seal, during these wrestling bouts between immature males, the penis may be unsheathed and copulation attempted. It is as if recognition of sex had not yet taken place.

Our picture of the young seals at the haul-out under Welsh cliffs resembles that reported of many gregarious animals which form herds and packs. Indeed one is tempted to bring in the society of man as an example again. In a large community of *homo sapiens* the various age layers are engaged in similar activities. The immature children play innocent games, tumbling and romping together openly. The teenage adolescents, more or less according to their home training, play together, less innocently, more sexually, 'necking' and fondling at the sweetheart stage. At or near puberty the young woman is concerned to secure the attentions of a male, often older than herself, to whom she has been attracted; but she does not readily permit coition, nor is he necessarily concerned to achieve this at first. Their libido is rising, and desire increases with each pleasurable exercise of loving contact and pre-play (special actions such as close embrace, kissing, and erotic pressure of genitals). There is of course the distinction that these preliminary intimacies are not usually conducted in the open assembly; yet where they can be seen at second hand, particularly on cinema and television screens, such observation of sexual behaviour is excitatory, increasing sexual desire in the young man and woman. There is plenty of evidence in the literature of both man and animal behaviour that sexual activity by one pair stimulates desire in other adults of the same species able to observe the same.

As the survival of the species depends on ovulation and conception in the virgin, it is obvious that all behaviour in the young animal must essentially be a preparation for successful mating. Even the principal activities of feeding, resting, and the avoidance of enemies in order to live are a preparation towards parenthood. But mating cannot normally be achieved without close intimacy. The ritual of breaking down the natural barriers of shyness, fear, and preparedness against attack with which all but the most sociable animals are endowed, either innately or by learning, must be observed.

In species normally solitary this is likely to be a slow and elaborate process. Even in gregarious species, which usually have a marked and closely observed order of dominance, the barrier of personal safety or individual distance, having been learned by a junior from tough physical experience of fighting or threats of fighting, is powerful and difficult to break through. We see this in the young grey seal, in the young human child, brought into contact for the first time with others of its kind. Innate or acquired fear and hostility are broken down quickly in the trained man child, and the sociable common seal, but far more slowly in the solitary-living grey seal child.

The assemblies of man and seal indirectly achieve the perpetuation of the species, both through the establishment of the dominance hierarchy which ensures order, and the temporary destruction of the same which enables the fertile mating of male with female. But anyone observing how the unmated adult (man or seal) does not normally permit the stranger to invade his individual or safety distance (in both this is at least twice the width of the body at its widest part, measured between face and face, in which member the signs of hostility or acceptance will be read) will realise how strong this invisible barrier is. Only in cases of stress and emotion can the barrier be broken: as when men and seals panic and flee in a tight mass at the sudden appearance of a dreaded enemy; also when man is forced to crowd temporarily in trams and tubes, and at football matches, cocktail parties, etc.

We have shown that in the grey seal assemblies the six-month-old pup, returning perhaps from a winter of wandering alone, has not broken down this barrier acquired from the example of its dam, nor will it normally do so until it has passed another year in the company of its kind. When the desire to play manifests itself, the youngster will play solitarily, with seaweed, driftwood, a glass float, and only gradually will it join in the games of more senior seals as it too becomes older and bolder.

The common seal pup behaves differently. At all ages from weaning onwards to its first mating it is playful and intimate with its contemporaries. We have described the playfulness of these pups (page 40). The Venables consider that there is a decline in the spring sexual play of rolling in pairs in the yearling, but it is evidently renewed in the sub-adult a year later, at two years of age, and until full puberty. Adult males, as in the grey seal, take part, though not often, in these water-dances in May time, but pregnant females never do.

On one occasion the Venables noted, in the Shetlands (June 2nd) 'some underwater play-chasing by a common and a grey seal, both sub-adult. Throughout our whole visit (May 23rd to June 10th) we saw periodic rolling by "pairs" of grey seals, and sometimes one of them had an extruded penis.' Evidently the water-dances of grey seals continue later in the Shetlands than they do in Wales, but this may be explained by the fact that when the inshore fishermen begin lobster-fishing in April, the assemblies in Wales are disturbed, and, for example, the largest of all – the Red Wilderness – is then dispersed altogether and until the fishermen haul up their boats in the autumn. Traditionally the seals have been shot and killed for centuries in Welsh waters, and although this persecution has of late years been much reduced, grey seals bolt to the water when a boat approaches.

The meaning of these water-dances or 'rollings' (as the Venables prefer to call them; they are certainly rollings in the surf, but are more like elaborate dances in the water) as observed by them, and by Selous in the Shetlands, and by us in Wales, seem to me plain: they are, we repeat, like the intimate dancing and social activities of sub-adult men and women, an essential preparation for successful mating. They remove the barriers of individual distance, of fear of attack. They excite, prepare and exercise the virgin penis which, before puberty in the seal, is constricted within a foreskin adhering to some extent like that of uncircumcised man. When its intromission in the virgin vagina, after many exploratory thrusts, is at last achieved, it may not result in a fertile mating, but it must prepare the virgin for her normal first impregnation (which, it is believed, takes place in the autumn in both species of British seals) by penetrating any hymenal obstruction of the vagina. While desire is pleasurable the first penetration in the mating of the seal, as in man, may give pain, and it is possible that this pain causes the virgin cow seal, at the moment of attempted intromission, to turn around and bite at the male's head – a frequent manœuvre in these sea dances. But whether penetration is achieved or not, the intimacy of close contact will prepare the virgin seal for the moment of fullest desire and oestrus at the autumn mating.

As to the precise age a seal must be before it becomes fertile and produces young there is some doubt. It was originally thought that if the domestic cow, and even more the huge whale, are mature sufficiently to conceive the first calf at two and certainly at three years of age, the cow seal, a much smaller mammal, might do the same. Laws found that in the South Georgia elephant seal virtually all cows mated at two, and produced a pup at three years of age. But subsequently Carrick and others found that in the same seal on Macquarie Island no cows pupped at two, a few did so at three, and the majority at four to six years of age. This is explained as probably due to the fact that the South Georgia population is exploited by man, who by killing a substantial quota each year but leaving a strong balance of breeders, chiefly sparing the cows, has removed much intrasexual pressure or competition for females and nursery territory, and so encouraged a more precocious maturity in a herd which has abundant food and ample space at the breeding grounds. It is known that bulls come into breeding condition earlier at South Georgia than at Macquarie.

It has been shown by Rand that all virgin females of the exploited southern fur seal, *Arctocephalus pusillus*, are mated at the end of their second year. But in other seals with exploited or stable populations it is reported that females rarely become pregnant before they are four or five years old, and males are not usually fertile before they are six or seven years old. These populations are perhaps subject to repressive behaviour of the adults towards the adolescents; in other words intrasexual competition is severe (compare with man!).

Harrison considers that the promiscuous common seal bull is not sexually mature before it is five or six years old. The polygamous grey bull may not be fertile until his sixth year, and may not secure a nursery territory until he is ten. As far as we know the females in both species are not sexually mature before their third year, and the majority one or two years later.

5 The Seals in Summer

With the approach of summer the lives of the seals seem, to land-bound man, almost perfect. The calm green and blue seas are transparent. There is an abundance of young fish inshore. A warm sun shines for most of the days, encouraging idleness and sleep at each fall of the daylight tide, as the seals seek their basking places, bellies filled with food.

The grey seals complete their moult in late winter and spring, and enjoy a period of comparative rest from all save feeding activities throughout the warm days of summer. The virgin males and females continue to be playful, but the old breeding seals are sleepy when ashore. They are putting on a thick blubber beneath their new coats, in preparation for the nursery season when they will fast.

The mature cow, after the strain of feeding the pup and fasting herself in the last autumn, has lost about one-quarter of her normal weight; and although mated and pregnant again within three weeks of bearing her pup, she is not in a condition to nourish a growing embryo. The lactation period over, she deserted her pup altogether, and after a short period of feeding began her annual moult. Until this was complete she was still unfit to carry a developing foetus. In fact, over this period the fertilised ovum does not develop into a foetus. The device known as delayed implantation comes to her aid. The fertile egg within the uterus remains alive, but unattached and without nourishment, in the form of a minute blastocyst, sperm and ovum conjoined, while the cow recuperates and gains weight by feeding heavily. By January she is fat again, but her coat now falls out, and she rests a good deal in the assembly, sleeping heavily, perhaps for days at a time. Her moult is completed before the end of February. She is no longer under a physical strain. Feeding generously again she attains the pink of condition, and it is at this moment, at least three months after impregnation, that a true pregnancy begins. The hitherto free-moving blastocyst implants in the wall of the womb, about 250 days before it is born as a fur-covered pup.

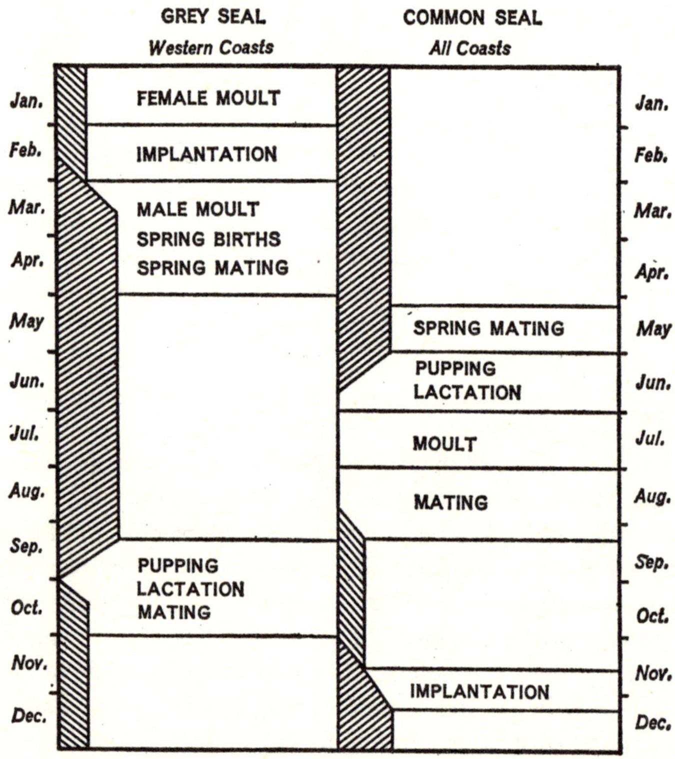

Figure 1. The annual breeding cycle of British seals. Periods indicated are approximate. Grey seals in the Farne Islands are born one to two months later; those in the Baltic and Canada four to six months later. The broad shaded bands indicate the period of true pregnancy, the narrower ones the false pregnancy period, i.e. the period of free blastocyst after fertile mating and before implantation.

It is about the time of implantation in early spring that the second matings described in the last chapter begin to occur. It is known that the moment of implantation is one of intense hormonal activity and high physical condition in the cow which produces secretions and enlargement of the uterus resembling that of oestrus. In effect implantation seems to induce a physical state resembling that at ovulation. It may well be that occasionally the free-moving blastocyst is lost from the uterus during the long winter quiescence, or dies; that spring matings of oestrus cows, described above, are a re-fertilisation which ensures pregnancy at the moment when the cow is able to support for the first time a growing embryo. But we do not know if these matings are indulged in by the cow already carrying a fertile blastocyst, or by the cow which has lost one, or even by a young cow which has never been impregnated before. Without killing her immediately after a successful spring copulation, and conducting a post mortem, it is impossible to know.

It has been suggested that a second mating at the moment of implantation might in some way help in the nourishment or reinvigoration of the hitherto dormant blastocyst; but it seems that, histologically, this is very unlikely. Delayed implantation after mating soon after parturition is a phenomenon known to occur in many mammals, particularly in wandering and pelagic species. It may be discovered in more. Its significance appears to lie in ensuring fertilisation at the one period during the year when the female is static while nursing her young, and therefore accessible to the male, which at that moment is sexually active and fertile. This would fit in with the picture we have of the grey seal's annual cycle, where it is now established that there is a delay in implantation of approximately 100 days after fertile mating. But it does not quite fit the case of the common seal, which mates after the moult, that is two months after parturition (one month after lactation ends), in August; but then the common seal is more sedentary than the grey, and the bull lives throughout the year with the cow in an established herd. The delay in implantation in the common seal is approximately sixty days (Figure 1).

We really do not know the reason for mating twice a year, or why a delay in implantation should occur. In some mammals delay is variable. There is one case of up to two years' delay recorded in the badger. But in almost every case of delayed implantation the full-term foetus is born into the outside world at the normal time for its species, which is usually the best time for its survival.

In the grey seal the usual time of birth varies with the locality (this is discussed on page 150). Yet, although the majority of grey seals in the British Isles are born in the autumn, there are records, at least in Pembrokeshire colonies, of births in every month of the year, with quite a spate, in some years, of births in January (5 records), February (3), March (6), April (10). It is suggested that these spring births are the result of fertile matings, without delayed implantation and possibly of virgin females, 250 days earlier (true pregnancy), which would mean that the matings took place from May to August, May to July is a period when not only are bulls normally infertile (without spermatids), but have not been seen to mate at the assemblies; mature bulls would be fertile by August. Or else that there is exceptionally a much longer delay in implantation following the normal autumn mating. Finally it should be noted that in the Hawaiian monk seal there are births in every month of the year; perhaps in temperate and warm latitudes (as Hawaii and Pembrokeshire) there is less need to concentrate pupping into a few weeks, as in Polar seals?

One observer, J. L. Davies, considers that the long-drawn-out breeding season in Wales may be due to lack of room on the nursery beaches, and the consequent need to stagger the pupping period (spring births are recorded in a similar cliff-bound terrain in the stormy Faeroe Islands, according to E. A. Smith). Incidentally these spring pups in Pembrokeshire form less than 1 per cent of the total pups born there annually; yet they are born at the same time as grey seal pups are born in wintry conditions in eastern Canada (page 152). One suggestion, that some of these spring-pupping cows have strayed across the North Atlantic, 2,000 or 3,000 miles, to pup on the western coast of Wales and the Faeroes, is hardly feasible since if it were so, one might expect some of the spring-pupping cows from the grey herd of the Baltic to wander to the Farne Islands on our east coast, and pup there in the spring, but Grace Hickling informs me that there are no records of spring births in the large herd there.

All through the summer the majority of grey seals of all ages live a comfortable life of feeding and sleeping. Daily routine is governed by tidal rhythm; their main feeding period is over high tide, and chiefly before it, when their fish food is also on the move, swimming inshore as the rising water floods the coastal rocks and flats. As the tide ebbs the seals seek favourite resting places on outlying rocks or sandbanks, on remote beaches either unvisited by man or where they have learned that man, although a visitor, is but a harmless spectator, a seal-gazing tourist (Farne Islands).

They lie out on calm days, enjoying the experience of drying fur in sun, but wind and rain do not prevent them landing – only a heavy sea will deter them. They sleep for hours if undisturbed and until the tide washes them into the sea. Some, perhaps those still not hungry enough to desire to feed, may remain over high water of neap tides when the haul-out ground is not completely covered. This applies to both species of seal. We have observed twenty-four-hour periods of rest in the grey; and the Venables record it in the common seal.

On still clear days at midsummer when perhaps the heat of the sun is trapped by rock face or sandbank, the fur-covered seal may become overheated. On such days our diary records that individual seals appear to find the heat stifling, and, slipping off the sun-baked land, they go to sleep on the surface of the calm sea, or at the bottom of the shallow inshore water.

In rough weather, when the basking sites are swept by heavy seas, and the surf makes it difficult for the seals to haul ashore on rocks, they do the same – sleep on the floor of the sea near by, just out of range of disturbance from the swell. From time to time – at intervals of from five to ten minutes – the sleeping seals must rise to the surface to breathe, but this is an automatic process and can be accomplished without waking.

Sun and wind are not essential ingredients for the full-fed sleepy seal to haul out and rest. Freedom from interference by man is of even more importance. Thus I have found grey seals regularly haul out in the stillness and darkness of beaches of shingle or sand lying above normal high tide where no sun or wind can penetrate. When marking adult seals in Wales at any time of year we could be sure of capturing some adults as well as sub-adults in nets laid at the entrance to certain caves which we could enter at low water, always provided we did not visit these caves too often (say not more than once a month) and so make the seals wary of using them as hiding places. Even caves which are completely filled at high tide would be used, if there was a beach or flat shelf to lie on during the lower half of the tide.

The fur of seals at rest in dark damp caves takes longer to dry out, but as the surface temperature of the body is high – it is warm to the human hand – it does so within an hour of coming ashore. In the light of our torch the cave seals would sometimes appear quite white and ghost-like at a distance, the huge eyes reflecting redly the artificial light, the whole effect being sinister in the extreme to the human entering the blackness of the cave for the first time.

The observer who is able to visit these hauling-out grounds, especially the pebble beaches used by a grey seal herd, will be conscious of the powerful, oily, fishy odour clinging to the stones; in the confines of an airless cave this smell can be almost overpoweringly pungent and sickening.

Given the necessary freedom from disturbance there is no doubt that some grey and common seals will haul out regularly each day all the year round and lie for hours at rest. Harrison Matthews has suggested that metabolic functions are drastically diminished during activities under water, and that long rest on land or at the surface of the sea is valuable as the occasion for digestion. The keepers at the isolated Smalls Lighthouse far out in St George's Channel off south-west Wales, have kept records for me showing the numbers of grey seals hauled out daily on the extensive tidal reefs which surround that lighthouse at low water. Here they are rarely disturbed by man, but are forced to leave the rocks altogether at high spring tides, which wash the base of the tower itself These records show that seals haul out on calm days all the year round, although their numbers are greatest in the long spells of fine weather in summer. The same regularity attends the hauled-out herds of the common seal in the Wash, which are at their highest numbers in the autumn when the new crop of pups mingles with the moulting or mating adults.

Common seals, frequenting estuaries and sandbanks, have somewhat better opportunities to haul out comfortably on the smooth surfaces of sand and mud. Even in stormy weather there is usually temporary shelter at low tide under the lee of a tidal bank where they can get ashore easily. But the grey seals which live on rocky coasts exposed to the huge rollers of the Atlantic gales are obliged to be at sea in the long winter storms or the short, sharp blows of summer; they find relief by drifting 'in the cradle of the storm-waves', or at the bottom of the sea.

Grey seals haul out in greatest numbers during two main periods: the pupping season, September to November in the west; and over the moult: January to March. Little is known about the moulting haul-out of grey seals in Canada and the Baltic, where the pups are born in the early spring or late winter, but presumably the moult must take place after the breeding season, and therefore in the summer, when these seas are free of ice.

The observer fortunate enough to study at close range a herd of a hundred or more seals, at rest and undisturbed on sandbank, beach or reef, will note that each adult lies separated from its neighbour by the usual safety distance, already quoted, of about twice the body's width (three to four feet) between head and head, and that no part of its body touches that of a neighbour, or rarely – as when the hind flippers are accidentally moved in sleep or sleepy movements and come to rest against the lower part of a neighbour's body. In that case, if the neighbour is fast asleep, or both seals are very sleepy, this slight contact of the lower parts of the body may be tolerated, perhaps undetected.

Usually however, any restless movement by one seal resulting in such contact is at once objected to by the neighbour affected: a flipper may be waved in the air towards the trespasser, and usually a hiss or a moan accompanies the movement. This might be sufficient to cause the offending seal to withdraw. If not, it may, perhaps considering that the other is the offending disturber of the peace, flipper more furiously at the other and wail. These duets are a feature of seal society ashore, particularly of the less sociable grey seal, but also of the common seal.

The wail of the grey on these occasions becomes a long musical lamentation – as heard at a distance, human-like, resembling indeed the plaint of mourners I once heard at an Irish 'wake'. This lamentation as of a human in distress, this music coming from over the waves of the sea, has given rise to numerous legends and superstitions of seals as human beings under a spell, and to tales of singing mermaids and mermen.

'Faith, when you hear the seals singing of a fine night in the bay,' said one man to me in the Bantry Bay area of south-west Ireland, 'be sure you are in for a big storm.'

'Maybe because, when you hear the seals so clearly,' I suggested, 'the weather must be very still, the air very clear, and these conditions generally precede a good blow of wind?'

'Arrah, but the seals know it too! They do be cunning, clever beasts. I mind the night when three foolish boys went off fishing in the bay without telling their people they had gone. There was a sudden storm that night, and we could hear what we thought were the cries of the seals on the reefs in the bay, for it is the low water time they do be coming ashore with their moaning man-like music. It was only next day that we remembered that it was high tide that we did be hearing these cries, and they were not the song of the seals, but the screams of our own boys, with their boat capsized and them drowning and no one but the seals to help them from the death in the sea which came to them.'

'Why do you suppose seals sing?'

'And why not? They do be cunning, clever beasts. They do be half human. Many a man in the west has hunted and killed the seal, but for myself it would be against my feeling, like murdering a child. And many a seal hunter has repented the day he killed the seal, for it brings bad luck, and spells a man they do say, so he and his people must pay back to the seal-folk a death for a death. But small blame to the fisherman, when he sees a great hole in his nets cut away by the big seal, for shooting the creature, and taking a guinea for the snout to help pay for the damage.'

The moans and cries of seals are not songs, I fear. They are entirely the result of petty squabbles over living space ashore. Seals do not sing like birds. Lost pups may wail for their mothers, but adult seals are silent when contented, and only break out into moans and snarls when discomforted by the pressure of neighbours. In a large herd local disturbances are going on most of the time, as individuals roll over in sleep, or ease their bodies into new positions, as man turns over in bed. The herd is quietest at low water; it is then spread out more widely, since the maximum of ground is available at that state of tide and accidental contacts are fewer; the whole herd may be peacefully asleep. But as the tide rises the herd is gradually compressed towards the higher ground, there is an upwards shuffling movement as the lowermost seals seek to keep dry and enjoy their rest a little longer, involving more and more invasions of individual safety distances.

It is amusing to note that once really dry a seal often tries to remain so for as long as possible. It will curl up its toes around its tail, as the rising tide laps towards it; but this cannot suffice for long, and the next move of the sleepy beast is to hunch upwards – only to encounter the protests of the seal sleeping above.

More and more protests break out, often in waves of wails and yelps, since the upward movement ripples through the colony, and a general restlessness is established. Those nearest the sea are forced into the water by the resistance of those above them. The youngest, lowest in the social hierarchy, give way to the larger dominant individuals who, perhaps settled originally at a low level, now crash their way to higher drier territories. Other seals, too sleepy to move, allow the tide to creep up and overtake them while still drowsy; the common seal may be seen bent into the shape of a U, its trunk under water, its head and tail above, until the last moment; even when covered it may not move, but go to sleep under the surface. In the clear water of calm weather one may see a number of seals asleep under water near the assembly ground, at any state of tide.

Probably individual adults will seek regular hauling-out sites in an assembly ground. Fishermen have told me that certain old seals can usually be found resting on the same rock or skerry day after day over low tide – they even give these individuals names from the name of the rock: the Tuskar bull seal, the Anvil seal, etc. Like old men who like their armchair, a seal, it seems, likes its rock niche as it gets on in years; but the habit starts when the seal is quite young. Day after day I have noted the same seals, both adult and sub-adult, lying out in the same rock niches offshore of the Red Wilderness. On one occasion I watched a half-grown grey seal defend its established territory on an outlying rock against the advance of a large bull; fortunately for the young seal it was in an advantageous position, a slight fold in the rock well above the slippery platform on which the bull had landed. Looking down from its enthroned position the young seal hissed vigorously at the bull, with all the moral right of an established proprietor; and after failing to convince the youngster, by prodigious yawns exhibiting formidable teeth, that it must clear off, the bull slid resignedly back into the sea, and went to sleep under the surface.

At sea such impudence by a youngster towards an old bull would not have been attempted; the sub-adults keep well out of range of the mature seal's reach. The same social order is observed at sea, as seen by me from the cliff-top. Except when playing their spring-time games the seals maintain the same respect for individual distance, the younger members swimming or diving away when within a few feet of an adult.

As seals grow old they become less playful, less tolerant of the younger members of the community, demanding a greater safety distance around them. Like crotchety old men and women, they become somewhat antisocial, readier to snap and snarl than formerly. This tendency is observed in many animals in captivity, including seals, sea-lions, and dolphins, which, amiable and easily taught to play when young, become unreliable and dangerous to handle from middle to old age.

We may take leave of this chapter by quoting the remarks of a pioneer observer of seals, Edmund Selous. In his *Bird Watcher in the Shetlands* he describes the typical behaviour of a bull seal as the tide ebbed and left him drying upon his favourite resting rock:

'The way he expatiates upon [his rock] it is delightful to see. Such great yawns, such stretchings, heavings and throwings back of the head, with supple curvings of the neck, such luxurious anticipation of repose to come – oh, such sleekness, such glistening. How intensely he enjoys this rest of his, his intertidal sleep. He was not asleep when we came (that would not have surprised me) but now as he lies at length, tippling a little, with the waves that tipple around him, the eyes begin to close, and even when he throws back his head and opens his eyes, as he does often, they may be shut or almost shut. Often he scratches his chin with his flipper or passes it indolently all over his face.'

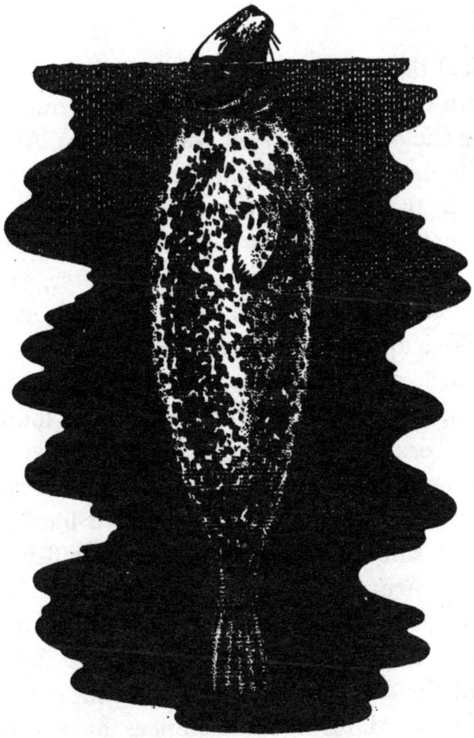

6 The Life Aquatic

The common seal throughout life is on the whole more playful, lively and sociable than the grey. Where they feel safe from enemies they can be seen enjoying the most exuberant water-dances. Although both seals will swim rapidly under water to gain speed for a leap into the air, the young grey seal in this sportive game only occasionally leaps right out of the water. Usually there is only a partial emergence, to the waist or tail, before the heavy grey body thuds to the water, to gain impetus for yet another breach. The common seal, however, regularly makes a clean dolphin-like leap, so that one may see the body as much as a foot clear of the surface; and returns to the water in a parabola head first, or straight as a projectile falling broadside with a loud splash.

The observer, from the cliff-top, has a perfect view on a calm sunlit day of the seal's mode of progress at high speed through the clear water. The fore-flippers are pressed tight, streamlined backwards against the body, to present a smooth surface. Propulsion is fish-wise, by a swimming lashing movement of the 'tail' half of the body. The hind flippers, with five toes joined by webbing are spread wide but pressed together so that they form in effect a vertical rudder or fish tail. With swift lateral thrusts of the supple thighs and hind flippers, the body of the seal can shoot forward for short distances at a dolphin-like speed. Dolphins are said to be able to keep up with a fast ship making thirty knots. But for the seal this would be a supreme effort over a very short distance, inspired by *joie de vivre* to gain impetus for the breaching leap; or by terror, as when escaping an enemy, including a rival seal; or in pursuit of fast swimming fish prey. Although we have never seen a seal attacked by its most dreaded natural enemy, the killer whale, other observers have reported the panic of a seal when a killer whale appears in the vicinity, and the tremendous burst of speed it will put on to reach the shore or shallow water, or to dodge this rapacious mammal if it is at sea.

No seal, in fact, has been seen to swim as fast as the dolphin, which normally, uninfluenced by the speed and excitement which a ship seems to induce, may be seen to slip along at around ten knots. (The famous Pelorus Jack, a Risso's Dolphin *Grampus griseus*, which accompanied the mail steamer regularly for over twenty years between North and South Islands, New Zealand, travelled at the bow of the ship, which did fifteen knots, for about six miles before frisking away on his private affairs; making his escort by night as well as by day; and receiving a special Order in Council protecting such dolphins in Cook Strait.) We have timed an adult seal moving swiftly along the coast towards a haul-out beach. There was no obvious current to favour or retard it. It progressed by swimming close under the surface, rising at intervals for air but not then reducing speed visibly. Our timing, measured against the parallel coastline traversed, proved that it was swimming at 5.5 land miles an hour. The forelimbs were flush with the body, and only the lateral thrust of the thighs and hind flippers, as already described, propelled its torpedo-shaped body. With its head thrust forwards, the neck was elongated, giving the seal a long lean appearance, otter-like.

When moving very slowly or when almost stationary the seal uses its fore-flippers as stabilisers, swinging them in short circles with a paddle-like motion, 'fanning' water towards the body. At the same time the hind flippers are vertical and separated from each other and swung from side to side with a slow twisting graceful motion of the thighs, each flipper opening and half-closing alternately. When proceeding at a moderate or normal speed the seal carries its fore-flippers close to the body but with the tips slightly projecting beneath, when their use as stabilisers (or 'lee-boards') is plain. The hind flippers are then lashed open and shut in wider swinging movements. Seals swim in this fashion equally well on their backs, and may survey the sea floor more conveniently this way, with their heads upside down – this I have often observed in both tame and wild seals.

Seals are all grace as they swim through their true element, water, but on land they move with an awkwardness that induces laughter and often pity or scorn. 'They're like slugs!' is a usual description. Even the more agile sea-lion, with its mobile hind limbs upon which it can progress by a shuffling waddle, raises a laugh at zoo or circus. But in the

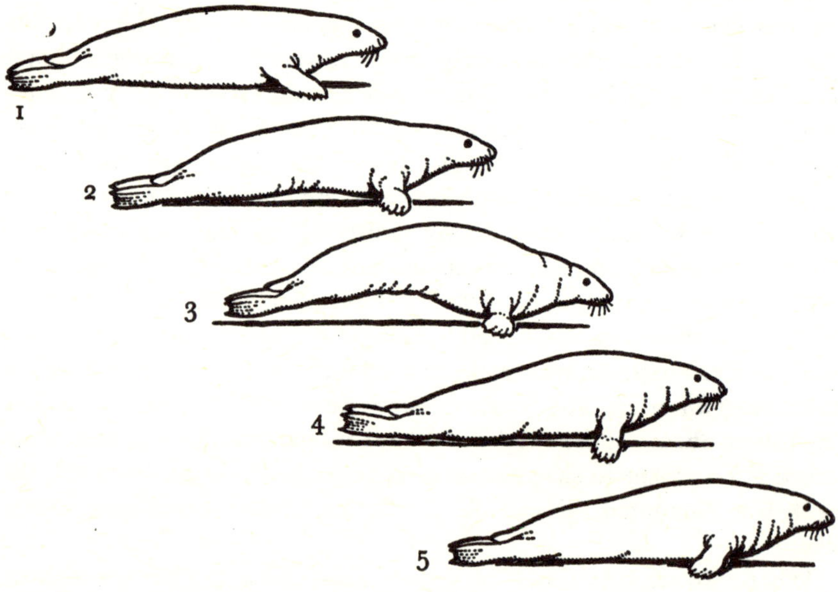

Figure 2. The method of progress of an adult seal travelling at speed on land. (When moving slowly, the adult may move its fore-flippers alternately, and like the young pup, drag itself along without any 'bounce'.)

1. Fore-flipper advanced and pressed to the ground.
2. Flexed fore-flipper takes the weight as the body rises, hind end drawn up.
3. The fore-parts are thrown forward by hind-parts on the lever of the 'hands', in a leapfrog motion.
4. The body collapses forwards as the lever movement ends with fore-flippers rearwards.
5. Momentarily at rest, before position 1 is resumed, with the fore-flipper about to move forward for a fresh grip of the ground.

phocid seals the rear limbs have become specialised and tail-like, and must be dragged along painfully and slowly by levering the body forward, either in a series of rowing movements, or, more rarely, one hand after another alternately. The head and upper part of the body is raised and levered forward by pressing the hands, fingers and claws clenched or flat, into the ground (Figure 2, page 74). The body is then levered until the breast touches the ground and, the weight thus taken off the front flippers, these are rowed forward for a fresh grip. Progress is slow uphill; but downhill the heavy body can be projected forward much faster. When alarmed a seal fairly bounces towards the sea over a smooth slope such as sand, pebbles and grass – faster than a man can walk. Daylight can be seen between body and ground at each violent bounce. But in rocky ground, amid large boulders, a seal makes but slow and painful progress, and is easily overtaken by an agile man.

When badly wounded from any cause, and losing the use of a fore-limb, a seal will not be deterred from coming ashore. Indeed it is probable that the wounds may cause a desire to rest on land. I have seen seals with a swollen, broken or otherwise useless fore-flipper haul out upon a beach without using the affected limb. Slowly but surely the seal will inch itself along: lying on the side of its rump opposite to that with the damaged limb, which is thus uppermost, it raises head and tail to form a U, and then gives a little leap or shove forward (the reader should try to get along in this position – it's not easy!).

Each of the five digits on each of the four limbs has a strong, narrow claw, one or two inches long according to age, and bluntly pointed. Those on the forelimbs are useful for gripping anything and everything, from the ground to fish prey, or another seal; they are longer than the hind claws, and are used, in conjunction with the very mobile arm, for scratching and preening almost any part of the body. There is no thumb: the five fingers with their claws cannot be used or flexed separately, as a man uses his fingers; they lie close together, like those of man, and are moved – opened, closed or clenched – as one instrument.

The hind flippers are even more specialised. They cannot grip or hold anything. They are relics of earlier terrestrial days when the seal (as the skeleton shows – see illustration on page 76) moved on four articulate legs like the otter (one species of otter however is already clawless – an interesting example of a mutation having sufficient evolutionary value to become dominant through natural selection).

The webbing between fingers and toes has developed to provide a broad fin-like surface for swimming purposes; in the grey and common and some other species this webbing extends beyond the claws, which are useful as an extension to the long bones of the fingers and toes, supporting the webbing like the ribs of an umbrella when it is extended on the propulsion stroke in swimming. Claws and webbing, fingers and toes of both hands and feet are folded, fist-wise, on the 'exhaust' stroke.

While the forelimb of the seal (Figure 3) is efficient as a paddle for

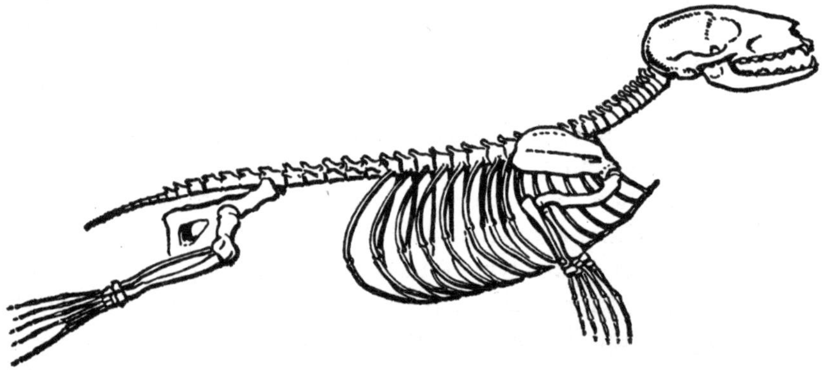

Figure 3. The skeleton of a common seal.

slow swimming motions and as a flexible hand for grooming and grasping motions it is comparatively small and unspecialised, and does not yield a broad surface for swimming like the hind limb. In the latter (Figure 3) the first and fifth digits, corresponding to the human big and little toes, are long and stout to support the arc of the fleshy swimming web. The second and fourth toes are shorter and thinner, and the centre toe is still shorter and thinner. Fully spread, the toes resemble a wide fan, efficient both as a steering rudder and fish-tail propellor.

The forearm is attached to the skeleton without a shoulder- or collar-bone, so that when a seal is swimming at speed, it lies flush with the body wall in the armpit or axillary hollow, and presents little resistance to the water, the tips of the flipper projecting, as already described, to act as stabiliser to the otherwise smooth keel-less body.

Seals can swim forwards or backwards, but are rarely seen 'back-pedalling' – as when retreating from the advance of another seal (especially an amorous bull from an angry cow), or when backing out from a narrow cave or cul-de-sac in the rocks.

Seals are perfectly at ease in the water, floating at the surface, diving beneath, gliding along at any angle, often on one side, or upside down. They can sleep floating or at the bottom, with a facility that arouses my curiosity and envy. If a seal is asleep at the surface, it will either float vertically, with its head above and its body perfectly upright; or it will float horizontally with the hump of its back only above the surface, the head and 'tail' submerged. To sustain this buoyant position without swimming movement to keep afloat, the seal must have a sufficient quantity of air in its lungs. Conversely in order to remain inactive and asleep on the floor of the sea, the seal must reduce its buoyancy – that is the amount of air in its body. This it appears to be able to do automatically, unconsciously.

I have often observed the sleep of grey seals under water from some hidden watchpost on a cliff above; and I have studied it in both seals in captivity. On reaching the surface after a sleep below, the seal immediately opens its nostrils, wide and round for a second and a half, or two seconds, exhaling stale air and inhaling ozone. The nostrils close for a slightly longer period before opening again for a similar spell. After about a dozen (varies from ten to fifteen) such openings and closings of the nostrils, the seal, its eyes usually fast shut in sleep all this time, turns and sinks head first to the bottom.

The duration of the period between submergence in sleep and emergence to breathe varies. On forty-five occasions in seven adult cow and bull grey seals the period spent below varied between four and nine (once eleven) minutes and averaged just under six minutes. Other observers have recorded longer periods. R. J. Harrison noted forty-five minutes' submergence for the common seal on one occasion; this must be unusual, since the usual period in this species is seven to ten minutes for adults, two to four minutes for weaned pups.

As the seal sinks in sleep, a few bubbles of air escape from the nostrils, probably the result of the increasing water pressure on the body forcing air from the lungs. Now, as the body on the floor of the sea must have a lower specific gravity than sea-water, the seal cannot rise to the surface without some effort of swimming upwards. But this it is able to do *while still appearing to be asleep*. Like man turning over in sleep the seal moves its limbs languidly, slowly, beating one or both fore-flippers, and twisting its 'tail' end a little, until it breaches the surface, head first. It may not be truly asleep, but its eyes remain shut above water. More effort, indeed, seems to be involved at times in the return to the bottom; which is either accomplished with a generous wave or scoop of one fore-flipper, or more rarely the sleeping animal drifts without effort to the bottom, back to its couch in the seaweeds.

How deep do British seals dive? It is doubtful if they go to any profound depth, but observations of seals in deep water are impossible to make without suitable technical aids. Experimentally common seals have survived enforced dives of fifty fathoms (300 feet), and can remain below for at least thirty minutes under artifical conditions; this seems to be the limit of their endurance in laboratory tests. Fishermen assert that they have caught seals in nets below this depth, but, like the story of gannets said to have been caught in nets at twenty fathoms, these reports may be based on faulty observation. Gannets do not dive deep, they are too buoyant with subcutaneous air sacs; but they get caught in deep set nets when, attracted by the sight of fish trapped in the meshes as the nets are hauled and close to the surface, they plunge upon the fish and become themselves entangled. We can well believe that seals are caught in the same way in deep set nets and trawls, as these rise to the surface.

A number of scientists have investigated the diving of seals, and their anatomical modification which enables this air-breathing mammal to remain so long below. Scholander found that a captive bladdernose seal survived a forced descent to forty-one fathoms. Kenyon records sixty to eighty fathoms for Steller's sea-lion. The human diver, rising suddenly from these depths – 360–480 feet – would collapse and probably die from the 'bends' or diving sickness, due to the blood circulation becoming obstructed with bubbles of gas – hence the use of decompression chambers on deep diving expeditions. Irreparable damage to the brain is suffered if man's supply of circulating oxygen is cut off for longer than a few minutes. But we have seen that the submerged seal remains below for periods and at depths corresponding to those recorded for the whale and dolphin. The record seal dive seems to be held by the Weddel seal in the Antarctic. Dr Kooyman fixed depth gauges on Weddel seals which recorded descents to 1,500 feet (250 fathoms) under the ice, a depth at which the pressure of the water is 700 lb per square inch or forty-six times that of the surface atmospheric pressure. One Weddel seal remained submerged for twenty-eight minutes in descending, and later ascending from such a dive – if the whole period was spent in diving and coming up again, the rate of travel through the water would be roughly one foot per second, but allowing for a feeding period of half that time below, this moderate speed is doubled. No one yet knows how fast a seal descends, or ascends, however, from such a depth.

Experiments with seals strapped to boards and plunged into tanks of water has shown that the heartbeat slows immediately from the normal 120 beats per minute when above water to ten, even to six beats per minute so long as the seal remains submerged. On coming to the surface the heart of the seal begins pumping at an accelerated rate, up to 160 beats per minute at first. This same slowing down of the heartbeat, though less pronounced, occurs in man when he holds his breath when under the water. But in the seals the blood-flow through the heart is regulated by a unique sphincter muscle near the heart, which reduces the movement of blood through the kidneys, intestines and extremities of the body – the flippers. These become starved of blood while the vital fortress of the body – heart, brain and lungs – continues to be fed with blood and its oxygen. As a result of this starvation the extremities of the body use up the limited amount of available oxygen, then switch over to the glycogen or starch stored in the muscles, which the body can burn up without oxygen. Prolonged diving with the circulatory system slowed down in this way, results in a reduction of temperature in the extremities. Seals have been found to shiver violently when brought to the surface after long submersion under artificial conditions. From such experiments it is discovered that seals regain their body heat rapidly as soon as they start breathing air again, whether ashore or floating at the surface. One can therefore suppose that the seal sleeping naturally for five or more minutes at the bottom of the sea receives the physical danger signals of cooling flippers and other de-oxygenation discomforts towards the end of each period below, and these automatically cause it to rise languidly to the surface to breathe. A seal under deep anaesthesia executes swimming movements automatically when placed under water, but ceases to move when placed on land again.

The amount of blood in seals and whales is very great in proportion to their size. Harrison and Tomlinson found the blood volume in a common seal about half as much again compared with the blood volume of a human adult. The body of the seal, when dry out of water, is warm to the touch, due to the numerous veins below the skin which intersect and cover the surface layer of flesh and blubber; the presence of these veins close to the surface results in profuse bleeding if the animal is wounded, or even cut superficially. We have noticed that the hind flipper of the adult is almost hot to the touch on the coldest day, due to this rapid circulation of blood at about 100 degrees Fahrenheit.

There are other adaptations to the life aquatic which are of great interest, but also not understood by man. How do the Weddel seals catch fish in the darkness of great depths below the ice, also in the total darkness which must reign there in the long night of the four months Antarctic winter? Do the Antarctic deep-water fish have luminescent spots? But one might as well ask how do seals find their fish food in dim-lit water, such as the shallow but opaque waters of muddy estuaries? The eye of the seal is large, larger than that of man. In clear shallow water I have often watched a seal hunting by sight, peering into rock crannies, and pushing aside the fronds of the inshore seaweeds, then darting after any fish or moving thing. In bright sunlight the pupil of the seal's eye is a pinpoint, but in semi-darkness it must be capable of full enlargement to catch the dimmest rays of light, and, although the sight of seal above water is probably inferior to ours, under water the seal's vision is likely to be acute. Above water a seal can spot a moving

ıto : P. Morris
lea Photographics

Common or Harbour Seal

Grey seals mating

Grey seal cow suckling her calf

Male and female common seals photographed together under water

Ardea Photographics

Face of a grey seal calf

Grey seal calf which lost its mother at birth. The body shrivels up during the week in which it may survive without food

Diana, an orphan common seal reared at Orielton, adored my secretary

Clipping a numbered ring on a three-week-old grey seal. An assistant stands in front to discourage the youngster from escaping, and to distract its attention from the ringer – not very successfully in this case

Grey seal nursery, Orkney. The cows have fled on the arrival of human observers

Grey seals basking on the beach at the Red Wilderness; note that each individual keeps a clear space between self and neighbour

Master bull in the centre of his harem of cows, with calves. An inland colony, Orkney

Master bull with a dozen cows herd together unnaturally in a panic movement towards the sea as the photographer disturbs them at an inland colony of grey seals

Photo: Anthony & Elizabeth Bomford
Ardea Photographics

Grey Seal

The Red Wilderness from the cliff top

Grey seal cow on left, showing small head, long neck, position of teats low down on body, hind flippers folded in resting position. The bull on the right is darker. Note thick, scarred neck and heavy head with large snout. Taken in October after the breeding season, when the cow is in lean condition

Activity off a grey seal nursery beach. In the foreground a cow looks for her pup. In the water to the right a bull holds a cow by the nape during copulation while another cow looks on

Common Harbour

Photo: K.W. Fink
Ardea Photographics

Common Seal, underwater

Photo: Anthony & Elizabeth Bomford
Ardea Photographics

human figure at a hundred yards, but does not appear to recognise man if he remains unmoving within twenty yards unless he is silhouetted against the sky. He is more likely to hear and scent an unmoving man at short distance, for both these senses are quite acute above water. If we examine closely the eye of a seal we find the front layer of the iris is transparent and traversed by a network of blood-vessels, which doubtless assist the eye in underwater use. The eye-ball is lubricated by an oily grease or tears which continually flow from a duct in the inner corner of the eye; in man this duct is connected with the nose, and our lubricatory tears flow internally – unless we are moved emotionally to copious weeping, when they flow both ways. But the seal has no naso-lachrymal duct, which explains why, out of the water, seal cheeks are always wet with tears – and these tears can flow more copiously, as in man, when the seal is alarmed, frightened or otherwise emotionally agitated.

Despite these adaptations to seeing under water, a wild seal which is totally blind can feed successfully, and thrive and breed normally (see page 168). So obviously sight is not essential. As to smell, the nostrils must be tight closed under water to exclude water from the lungs; and in fact they are never seen to be open below the surface. This does not altogether preclude seals from smelling under water – scents might be filtered from the water past the oily film which excludes liquid entering the closed nostrils – but there is no proof of this. Then does the seal hear its prey moving under water? But the tiny ears of the common and grey seals are automatically closed by muscular action under water, and any sounds must be fairly muffled as a result; but again we do not know enough about seal hearing below water. There is only the sense of touch left, or some other sense we do not know of.

Any movement under water sets up vibrations and currents which must be felt by the seal. The location of the centre of the disturbance is conveyed – in the absence of visual aid – to the seal's brain through the long stiff whiskers or vibrissae attached to the snout and brows. These antennae are continually growing, highly tactile and sensitive, and connected to the brain by substantial nerves. The upper lips of the seal (see photograph, plate 3) form a thick pad with a positive bouquet of a score of such whiskers deeply embedded on each side. When questing under water, the seal can thrust this pad in a sweeping movement to and fro, by pushing the highly mobile upper lip in and out; the whiskers are stiff enough to stand out clear of the face under water, except when the seal is moving at high speed, when they trail back along the cheeks and neck. The sense of touch seems to be the main sense employed to catch fish prey in darkness below the surface. Incidentally, seals use their whiskers involuntarily when out of the water. Orphan common seal pups would greet me by extending their lips and whiskers, as if expecting a kiss, and then when I extended a finger to let them smell me (they knew me well by sight as well as scent), would snort gently as if to blow out an unpleasant smell; in reality they were not expecting a kiss or rejecting my scent – it was the traditional form of greeting of the common seal, the snort being one of recognition perhaps.

Having caught fish food under water, how does the seal deal with it? The sharp pointed teeth of the phocid seal are adapted solely for catching, tearing and swallowing, as opposed to holding and chewing. For this reason they have fewer teeth than other carnivores which chew their food. The formulae for the adult grey and common seals is:

	incisors	canines	postcanines (molars)	total
Upper Jaw	6	2	10	=34
Lower Jaw	4	2	10	

(The adult teeth are often already showing at birth. The deciduous (milk) teeth of the pup are usually lost before birth; they contain fewer postcanines – 12 instead of 20.)

These teeth deal efficiently with a wide range of fish-foods. The molars of the common seal are cusped for crushing shell and crustacean and mollusc. Those of the grey seal have one large and two small points for gripping fish. The long canines come into play first, to catch the fish. If small the fish is turned round and swallowed head first, whole, and it could be still alive. If too large to make one mouthful it may be brought to the surface, held in the fore-flippers, and torn with the teeth into smaller chunks for swallowing. According to the availability of fish locally, the diet of the two British seals is wide and opportunistic. The contents of about 200 stomachs of the grey seal for which information has been published have included predominately fish of many species, including salmon and sea-trout, and many remains of cephalapods. The latter – squid, octopus and cuttlefish – are generally unpopular with our fishermen because they are unsaleable on British markets, but also because they attack and devour saleable crustaceans such as lobsters, crayfish and crabs (seals and whales are the only known predators of cephalapods, as far as I know). Some Orkney lobster fishermen have told me that they welcome the the presence of seals near their pots, for then the cuttle and squid disappear.

Seals have very rapid digestions, and normally soft fish pass through the stomach within a few hours; but the beaks of cephalapods, indigestible, tend to remain in the stomach long after all soft parts have been assimilated, and may account for an undue emphasis on the proportion in the total diet.

The contents of 100 stomachs of the common seal on the Pacific coast of North America comprised 93·6 per cent fish of more than ten species, crustaceans 5·8 per cent, and molluscs 0·6 per cent; salmon were found in only two stomachs. In twenty adult common seals taken in the Fraser and Skeena salmon fishery rivers on the same coast salmon formed 28·5 per cent of the contents of five stomachs only.

In the Wash area common seals, in addition to shell-fish, shrimps, prawns and molluscs, take practically all local species of fish: whiting, goby, dab, plaice, flounder and other flatfish.

In Welsh waters grey seals seem to have an omnivorous fish diet. I have watched them bring large fish to the surface to kill and eat, chiefly skate, ray and conger eel. Sometimes the fish has been almost as long as the seal itself, if its long tail is included. I have several times seen a seal wrestling with conger or ray. A large ray is tightly held in teeth and fore-flippers as the seal surfaces. The seal begins its meal upon the still wriggling, slashing fish by manipulating it until it can grasp the soft white underside. A long strip is torn away, and swallowed with a jerking movement – a dog-like gulp. The maimed ray is then allowed to float free; it swims feebly, threshing its wide 'wings' downwards. The seal now dives and, almost playfully, like a cat with a mouse, brings it to the surface again, pushing it about with its paws. Having turned it over to expose the white belly again, it proceeds in leisurely fashion to tear away another strip of the white meat. Gulls hover overhead the blood-stained water, and dip to pick up the fishy crumbs during the seal's orgy. As soon as it is satiated the seal leaves the carcase, including the head, backbone, and bony tail, to the gulls, which endeavour to keep it at the surface by plucking at it in shallow dives – the best a gull can do. However it gradually sinks out of sight and range of the gulls, to be eaten by lobsters, crabs and other bottom-feeding scavangers.

A grey seal will treat a large conger in much the same way. It brings the eel to the surface, still wriggling wildly but evidently half-paralysed by the initial bites. Holding the eel in its fore-flippers the seal proceeds to bite away and swallow chunks of the solid flesh along the flanks of the eel from head to tail, much as a man chews off the corn from a maize cob; and leaving the tough head and the backbone intact.

Kindly visitors to see my orphan seals would sometimes be worried that they bathed in fresh instead of salt water; and I would have to point out that some seals live exclusively in fresh water. But evidently marine seals are adapted to ingesting a high concentration of salt. Our tame seals would drink clean salt water occasionally but not as freely as they would drink fresh water; they would suck up a few mouthfuls of fresh water after every feed of fish, provided the water was clean – they would not drink dirty water.

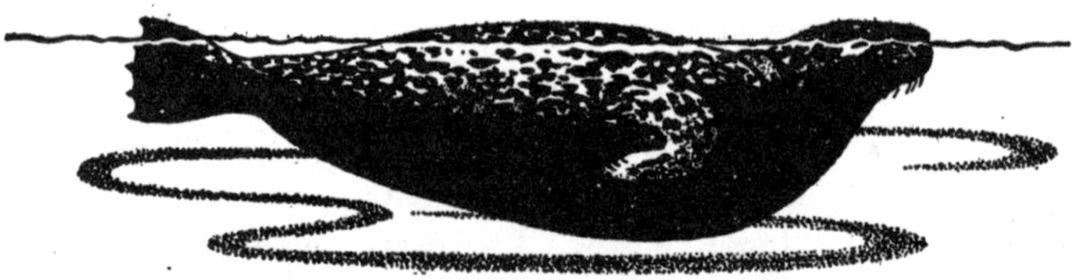

7 The Breeding Season

All the dark night a white swell had raked the island beach beside which our boat was anchored. We heard the claws of the surf rattling through tumbling pebbles. At rare intervals there were little silences as the sea rested, and sometimes these moments would be broken by a wail or a scattering of pebbles as a seal made voice or movement.

Sheltered from the Atlantic by a huge cliff facing north, we lay off a steep beach about fifty yards wide, on which grey seals established a nursery each autumn. But it had been too dark for us to note more than the dimly seen and heard movements suggesting that seal activity continued by night as by day.

The middle days of September on the far western coast of Wales. Under high cliffs, through violent tide-races, in and out of caves, on islands, islets and mainland haunts we had lived with the seals. Today the sea grew calm as the sun parted the morning clouds, and glittered upon the drops of dew on furled sail and pristine cliff-turf, and lit up the wet faces of seals swimming around the little yacht.

The pebble beach off which we lay at anchor was part of an island sanctuary, and typical of beaches on the two main breeding islands of the grey seal in Welsh waters – Skomer and Ramsey Islands. That day there would be an exceptionally low, equinoctial spring tide, most favourable for seal watching, and for entering caves exposed only at low water. Taking the dinghy ashore I climbed to a comfortable niche in the heathery cliffs on the south side of the island where the sun was already delightfully hot. I opened my notebook to record the events on this nursery beach, as the tide rapidly receded. What I recorded, and put down here, is typical of the behaviour of the little colonies of breeding seals on this coast.

Wherever one finds a healthy young white-coated calf on a beach one can look at once for its dam, and next for the bull which will always be waiting upon the nurse cow. The bull I could see was a large black-brown beast, probably a full ten years old, with thick wrinkled and scarred neck, and heavy convex roman-profile head and blunt nose.

Each mature bull acquires by right of might a nursery territory, which may consist of a small narrow beach a few yards wide under a cliff or in a cave or upon an open islet, upon which one or more cows have deposited their pups; or of a wider beach with two or more cows and pups. Still wider beaches with larger numbers of cows are divided between several bulls, and there may be considerable skirmishing and sometimes severe fighting where their territories meet. Dominance is maintained by the bull patrolling the water along the edge of his beach, and driving away other males. So long as he can do this he is a 'master' bull, and he will cover the nurse cows as they come into oestrus about the fourteenth day after parturition. This appears to be his sole interest – sexual, not paternal – in occupying a nursery territory.

Bulls without territory but sexually potent are constantly on the prowl to mate with oestrus cows, and will immediately enter a master bull's territory if the dominant bull does not prevent the trespass. Sometimes, especially after mating and perhaps during the long union of seal copulation, the master bull falls fast asleep, and is in no condition to notice a trespass by a swimming bull. In that case, as I have seen more than once, a bachelor or secondary bull might succeed in mating with a second cow which happens to be on heat at the same time as the first and ready to accept the attentions of a bull living on the fringe of her nursery and already known to her as such.

Conditions were ideal for observation at Skomer on this calm sunlit day. The master bull lazily parted the smooth water, making a chevron of oily ripples as he patrolled the seaward edge of his territory. The three cows and their calves were all asleep, the calves on land, the cows under the shallow water.

The bull passed over the sleeping forms of the cows, which from the height of my look-out on the cliff lay like pale silver-blue torpedoes at rest and laved about with the deep brown oar-weed fronds. Possibly the cows opened their eyes sleepily as the bull glided past, but I could not see this; I saw only that the cows accepted this close patrol as from

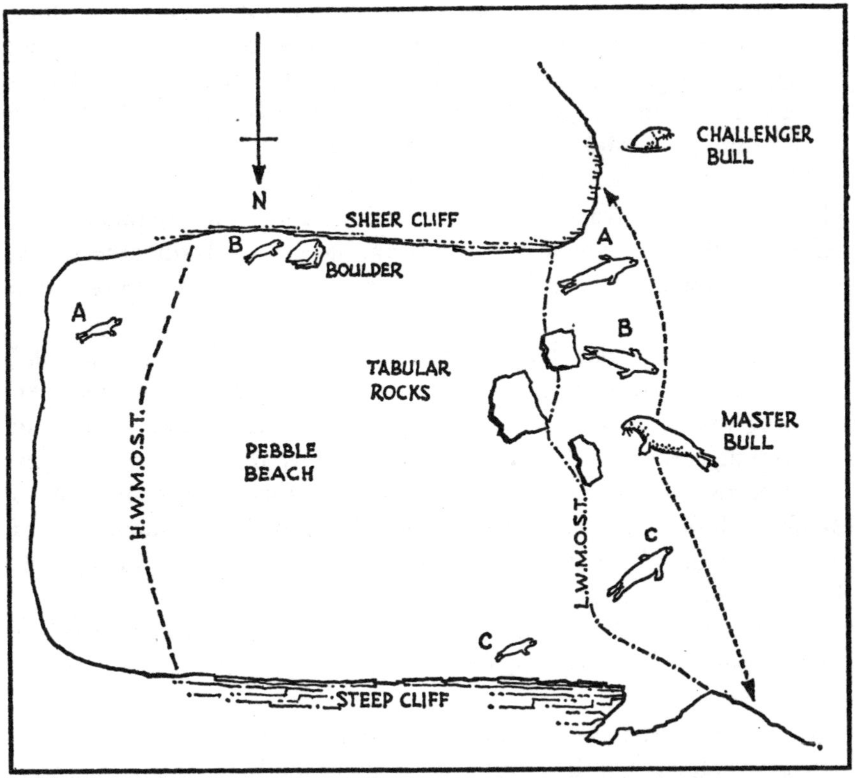

Figure 4 (not to scale). On the left, pups A, B and C. On the right, their mothers asleep under water. The normal patrol range of the master bull is shown. A challenger bull lurks to seaward, beyond the high cliff, ready to take the place of the master bull when an opportunity occurs.

habit; they must have sensed and even enjoyed the protection of the bull. As a sentinel the bull was performing a useful job – he would certainly sound the alarm and thus wake the cows if man or other enemy threatened the peace of the nursery.

There was little more movement until well after noon. The cows, asleep under water, rose at regular intervals to breathe. Through the telescope I could see their nostrils open and shut on an average a dozen times before each re-submergence. Even the bull seemed infected with the noon heat, and the torpor of his women. He halted for long moments, resting his head now above, now below, water, lying horizontally with his back a shining oval in the sun and air. Occasionally he pivoted, so that his head pointed heavenwards, and for a while he floated upright, blinking peacefully.

When one cow woke completely and swam to the beach, the bull followed her. She turned on him with a snarl and a wave of one flipper, her little tail erect above water. He swam back towards the other cows, obviously unwanted; courteously he did not press his attention upon her.

The falling tide had exposed some roughly tabular rocks around which the bull swam with difficulty, grounding for a moment, then splashing into deeper water. This slight disturbance woke another of the sleeping cows, herself becoming stranded by the ebbed water. She appeared to panic for a moment, and hustled through the shallows to attack the bull. She beat the surface with her flippers. He moved out of her way a short distance, his mouth opened in a prodigious yawn, half-threat, perhaps half-boredom.

The first cow seemed to study this performance, her round eyes just above water, her nose blowing noisy bubbles at the surface, like someone gargling.

It was time to stretch my legs and go aboard the yacht for lunch.

When I returned in the middle of the afternoon the oldest of the three calves (A in Figure 4) appeared to be suffering from the heat of the sun reflected from the stones of the beach. He began to moan, a wailing note: 'Soh-ra-ab!' He was half grey, well into his moult, and had for some time been restlessly rubbing his coat against some large pebbles, and scratching now with one, now with the other fore-flipper, as he moved or rolled about. Suddenly he began to hunch down the beach, making for the shadow of the cliff. As soon as he reached shade he ceased to move or moan.

A second calf (C), all white, also seemed troubled by the sun. It moaned in a gentler, more musical key, and shuffled about until it was in shade.

The third calf (B), also white, had been partly in the shade of the cliff already, and continued to sleep solidly except for an occasional flexing of the limbs, opening and closing the flippers, as a man might subconsciously stretch while at rest.

Presently the tide began to flow, rapidly sweeping round the tabular rocks again. The sun sank behind the mass of the island westwards. The cows were less sleepy. Frequently one would raise her head high, 'standing' on her hind flippers in shallow water, and gaze at the calves.

Suddenly calf B emerged from behind its boulder parasol and calling 'aye-ow!' at intervals, humped down towards the sea. One of the cows responded by standing on her tail high in the water as if to locate the noisy calf by sight; she then came ashore rapidly. The pair met halfway up the beach. The cow sniffed at the calf, and reassured as to its identity, rolled over on her side. Her teats were already protruding, before her calf touched her.

The calf nosed her belly, quickly finding the teats. For twelve minutes she sucked milk, changing from one teat to another every half-minute or so. As soon as the calf ceased to feed the cow moved down the beach, looking back as if to invite and wait for her calf to follow. It did so, rather languidly, being full of food, and with many short pauses. But once in the sea it began to play with its mother in a delightful manner. It mouthed her face, climbed over and under the huge female body, chased after her when she turned aside, and even managed to climb on to her back and steal a ride. This cow was a huge, magnificent animal, marked like a jaguar on throat and belly with diamond-shaped cream spots, and she had buff-coloured brows.

The play lasted about ten minutes, then the calf returned ashore and with business-like heavings of its rotund white body, it climbed up the beach to its former resting place beside the boulder. It fell asleep at once, a tired, full-fed and contented baby. After watching it intently for a while from her position in the shallow water, the cow slid under the flowing tide, and sank to rest in the oar-weed.

Meanwhile much the same procedure had been adopted by the other two cows. One, the pale-coated mother of the half-grey calf, had barely reached the edge of the sea before her child descended the beach at speed to meet her, and nudging her trunk, almost rolled its mother over in its impatience to feed, and took a teat in its pursed mouth. But judging by the rapidity with which the calf switched from one nipple to another, there did not seem to be an abundance of milk available. Moreover the cow's welcome had been comparatively indolent, lacking the solicitude of the first cow. From this reception, and the coat-colour of this calf we judged it to be at least two weeks old and near the end of the lactation period.

The third cow was so huge and dark, that she could easily have been mistaken for the bull, whose jet blackness in the evening light was quite equal to hers in depth of shade. She met her calf halfway. After it had fed she patted it with one flipper, then remained with it on the beach until night had fallen. Mother and child appeared to be asleep, peaceful, side by side, touching each other, blissfully oblivious of the others.

In the shallow water the bull had also come ashore during the haul-out of his three cows. But he did not proceed far up the beach, just a body's length, enough for his fur to become partially dry. He was attentive to the pale cow, mother of the half-grey calf, but in the gathering darkness I could see nothing clearly. There was a splashing and a wail from the cow, objecting to his courtship. I climbed out of my cliff niche, and crossed the island, wondering if supper would be ready aboard the yacht.

It was not until the next day that I witnessed the completion of the courtship. From the time I took up station on the cliffs I saw that the bull was more active than usual. He followed the pale cow ashore at 8 a.m, just as I arrived, but she beat him off while the half-moulted calf suckled her close to the water. The bull's presence so near, his eager lifting of his head and sniffing towards her, as if the peculiar odour of her approaching oestrous state maddened him, also distracted the cow.

She left the calf abruptly, and advanced on the bull with a loud musical hooing or hooting cry. With wide-opened mouth she snapped at him. He waved a flipper as if to ward off the bite, and suddenly threw his bulky body towards her in the shallow water, splashing heavily. She rolled over the intervening pebbles as his jaws grabbed towards her neck, and bounced into the surf.

He chased her in the sea for two hours, not violently, but steadily. He seemed to know that her resistance was weakening, that she had begun to be playful rather than belligerent. She swam from him under water, twisting between rocks in the weedy shallows. He followed her closely, frequently thrusting head and shoulders above water to breathe, and to gaze down upon her. Sometimes this undulating pursuit became quite fast, raising a white bow wave at the surface. Like the dance of the young seals in the spring, already described in this book, it was a frolic of love, but with a far more determined purpose.

At times he seemed to tire. The pace slowed down. They trod water a few feet apart, waving flippers in gentle fanning motions, heads at the surface, gazing this way and that as if pretending not to know each other. If he volunteered a nearer approach after this peaceful interlude she would make a snatch at his face with open jaws and musical moan.

Presently her vocal protests subsided. His persistence was breaking down her coquettish will to put off the moment of surrender. When she glided away he moved parallel or slightly behind her, above or under the clear blue water. This pausing and gliding away was repeated several times, circling the little nursery bay. As soon as he attempted to close upon her, she would turn sideways or make a somersault and swim in the opposite direction. Twice he anticipated her, and she fell into his outstretched arms beneath the surface, only to wriggle free. It was a charming dance, a device of love.

In succeeding mêlées there was no sound now save the splash and spray of their friendlier clashes. They were closing together. The cow was even turning around the bull in an affectionate gyration, her mouth but half-open, like a wayward playful bitch. She made no further attempt to leave him.

Just after 10 a.m. the movement of the love dance became more involved; and the bull succeeded in gripping the folds of skin on the cow's nape. Then his fore-flippers closed tightly about her diaphragm, below her arms, the long nails pressing deep into her skin. He lay along her back, 'tail' pressed close around hers. It was difficult to follow the moment of union as the genital regions were masked by the hind flippers and tail of the bull, as well as by the refraction of light through the sunlit water. But the hind quarters of the bull seemed to ease sideways as if to meet those of the cow, which were raised slightly, and her tail remained free and lifted.

The date was September 17th; the time 10.20 a.m. The bull relaxed his grip of the cow's nape, but held her firmly in his arms, and the pair remained united for forty-eight minutes, floating with eyes closed for short intervals between longer spells of resting at the bottom of the sea, only one fathom deep. When the pair rose to breathe it was the cow who used her fore-flippers to swim upwards, carrying the bull in coitus with her; the bull's flippers were engaged in holding her. His head now rested comfortably along her nape and his mouth was closed. When at last he released her she turned and brushed past him into deeper water. He floated at the surface sleepily, and she slept below the surface near by.

A coition of this length is usually followed by slumber, even though the pair have appeared to sleep during the long mating. In the book *Island of Skomer*, John Buxton records a similar full coition by the grey seal lasting over an hour, in the sea.

Copulation in the sea is a smooth, seemingly effortless operation for a mammal so essentially marine as the seal; and owing to its awkwardness on land, the seal mating ashore appears to be a clumsy manœuvre. One writer describes the phocid seal as the most ill-adapted mammal for terrestrial existence as to its locomotion and copulation. Yet the grey seal, in some of its largest colonies in Scotland, commonly mates on land.

I have witnessed matings both in the sea, in the surf, on the beach, and inland. The bull pursues the same course of following the oestrous cow about until he can grip her nape. In doing this he must hold her fast with his sharp canine teeth until he can wrap his fore-flippers around her and establish coitus. He inflicts small wounds on her neck with his teeth; usually these are insignificant, but I have seen a cow bleeding profusely after escaping from an amorous bull. Small permanent scars appear on the shoulders and lower nape of the cow after these mating manœuvres. The huge elephant seal bull has large sharp canine teeth for fighting other bulls, and these are inevitably in play when he seizes the little cow during copulation; these powerful canines leave an increasing number of permanent scars on the necks of mature matrons; and rarely the bull in this species has killed a cow in biting through the skull instead of the skin in mating.

When I returned after a meal on the yacht that afternoon it was to find the bull had engaged with the cow once more. He was holding the loose folds of the nape of her neck in his jaws, but his hind quarters floated free. Perhaps it was the end – but it looked more like the half-hearted beginning – of another coitus. It came to nothing. She twisted free with a sudden movement. He followed her rather languidly. They passed out of sight around the southern headland. Presently he returned alone, gliding slowly along his normal patrol stretch fronting the nursery beach.

The cow returned within five minutes and floated lightly inshore. She half-rose in the shallows and gazed towards her sleeping calf; then turned to look at the bull, whose black, wet and shining head cleaved the water in her direction. She seemed to be torn between desiring the bull, and going ashore to locate and feed her calf. But it was now low water, and the calf far away over the rough stones of the lower beach. She thrust into the water to meet the bull.

At that moment one of the two other cows rose up from sleeping under water and collided with the bull. She woke with a violent splashing movement, howled at the bull, rolled on her back and beat the surface with both flippers. She was in no mood to play with a mature male. But in a few more moments her own calf came down to the sea's edge, laboriously scrambling over large stones to reach and be suckled by her; and for the next hour the pair played endless games above and under water.

The bull and the oestrous cow meanwhile moved off into deeper water, and resumed their courtship, chasing and gliding around each other. The cow now seemed to be pursuing the bull, and he in no hurry to respond. I could not follow what happened to this pair, but doubtless mating occurred again. They disappeared past the edge of the headland. I remained in my niche to study the behaviour of the two remaining cows and their calves. Both cows fed and scratched and played prettily with their calves during the afternoon. And presently the oestrous cow returned and came ashore to feed her half-grey child.

As the tide rose and submerged the tabular rocks the master bull, who must have been sleeping around the corner, at the entrance to the nursery beach, glided low in the water and grounded on a smooth reach of pebbles where the sunlight still lay warmly. He pulled himself laboriously ashore, as he had done yesterday. He lay down as if worn out by his sexual adventures. He fell fast asleep on his right side. His fur rapidly dried out.

Some twenty minutes later I noticed the scarlet penis unsheathed and protruding about four inches. It pulsated irregularly – at five-, eight-, and three-second intervals – while the bull slept heavily in the declining sunlight.

Out to sea the roman profile of the head of another bull had appeared swimming quietly across the entrance to the nursery bay. But although the visitor raised his head to look at the four adults and three calves on the beach he did not come ashore. He was not so large as the master bull, yet I judged him large enough to be sexually mature, not less than six years of age perhaps. The master bull was too soundly asleep to notice the trespasser; but I thought it possible that the younger bull could scent the powerful rank odour given off by the sun-dried body of the older bull; I could smell this even at the top of the cliff 100 feet above him. If so the young bull would be warned of the presence of the dominant bull protecting the three cows, including the oestrous, but already mated, female.

Sleeping, mating and patrolling are the main activities of the master bull. I have never seen a mature bull feeding during the month or two while he is attending his females at the nursery; nor is there any record of food found in the stomach of a dominant bull of any species of seal at this moment.

The scene shifts to a small green island in the Orkneys, lying in the tireless tide race of Stronsay Firth – Muckle Green Holm, where up to 600 grey seal calves may be born during the autumn. The main pupping period is a month later than in Wales.

Here and on one or two other islets seldom visited by man at that time of year, the pattern of behaviour of breeding bull and cow is a little more spacious, more leisured, not troubled by considerations of tide and ocean storms to the same extent.

The big master bulls are first to establish themselves in the nurseries. Ready to breed first of all the males, they come ashore as early as late August. They are thickly covered with blubber beneath sleek summer coats, and ready for the long fast of the breeding season. So long as man is absent they behave uninhibitedly. Crossing the landing beach, they climb up on to the grass and shuffle well inland (on the Green Holms, on North Rona, Shillay and Little Pentland Skerry) perhaps as much as 300 yards inland, and almost that height in feet above sea-level. From the beach, or the rock platform on which they land, they follow well-established routes or trails to certain lying-out sites; and it is thought that the same mature bulls return to the same individual nursery positions. A favourite site is beside a pool, in which the adults may wallow at times. They will follow a stream, the lowest line in a valley, which becomes a regular trackway, in order to reach a remembered site. They climb quite steep slopes.

Each bull settles alone, apart, separated from the next by a score yards or more. They are lazy at first, and somnolent, ignoring each other, for there is as yet plenty of room for each to stake his claim to a few hundred square yards of ground. Most of these first arrivals show their maturity in their size and the scars they carry about the face, neck and body, proof of fighting in previous breeding seasons. They are all dominant 'master' males. The young bachelor bulls do not usually come ashore until the cows have begun to calve.

One or two, or up to five days before parturition the cow hauls ashore and moves up, probably each season to the same plot of ground or grass where she gave birth to a calf the year before. This return to the same nursery site is true at least of some cows I have watched on beaches in Wales. Inland she settles close to a bull, or more rarely the bull shifts his territory to where she is established. But the pregnant cow, while appearing to need the company of a dominant bull, and making considerable physical effort to reach his territory, and lie beside him, will not allow him to touch her yet. She hoots and flippers if he attempts familiarity; and he wisely takes the hint and leaves her alone. She behaves, however, in exactly the same way towards any other cow near her, exhibiting what some writers have termed 'jealous' behaviour. In fact she is merely maintaining the rule of individual distance by which, most of her life, she has protected herself from sudden attack.

When the calf is born on the inland grassy territory far from the sea the cow remains with it for the whole of the lactation period of two to three weeks, unless disturbed and forced to retreat to the sea by the arrival of man. When this occurs, as on the visit of a battue or killing party, about two-thirds of the cows and bulls make for the sea. But in certain nature reserves where seals have been strictly protected for several years (e.g. the Farne Islands, North Rona) and little disturbed, the arrival of an observation party, moving around unobtrusively, causes much less panic, and fewer seals abandon their territories. It is obvious that some of these individuals have learned that man does not normally harm them. His visits, benevolent or otherwise, are always harmful, however, and there is a long period after his departure before the adults are settled back in their breeding sites; and some casualties inevitably occur on each disturbance.

A few individual cows will yield to no visitor, seal or man. I have often admired these mothers who refuse to allow me to come between them and their calf, wheeling around their baby as I have circled it, and lunging towards me with open jaws and fierce snarls.

It is obviously an advantage for the cow to remain ashore with her calf throughout the nursing period. Although the inland-born calf has nothing to fear from the sea at first, the mother's solicitude is none the less great. Darling has described this affection which he observed at the colony on the plateau of North Rona. After the birth of the calf the cow rested for a quarter of an hour, taking very little notice of her child; and

'Then she offers it her two teats and within half an hour the calf is taking its first meal of milk which is ten times as rich in fat as cow's milk; she does not lick it at all, though she will smell it. Immediately after the first feed the cow begins to show maternal affection, which increases in intensity during the following three or four days. She shuffles round in order to touch the calf with her muzzle and then to scratch it lightly with her forepaw. This scratching is a habit invariably practised after feeding has taken place, and after the first feed the calf comes itself towards the mother's head and is scratched from head to tail down the back.'

The picture is idyllic, and as we have seen this delightful behaviour applies to each grey seal calf wherever it is born, but the calf born inland continues to enjoy this attention uninterrupted by any movement of the cow to and from the sea. In this environment of continuous maternal care, records Darling, the calf sucks the abundant milk every two hours or so, then gradually at longer intervals. Free from the dangers facing the calf born on a tidal beach, it scarcely moves from the spot where it was born, for the first two weeks.

As on tidal beach, so on inland turf, the mother is not afraid to attack the master bull if, in his mating manœuvres with neighbour cows, he rolls towards her calf. The bull is surprisingly well-behaved when attacked by an angry matron defending her child. He will either retreat or yawn at her threateningly, with a fore-flipper beating the air to keep her from actually biting him. Such flipper waving and moaning yawns are signals of appeasement and are usually sufficient to settle the dispute as he edges away. But, as I have said earlier, nothing will make him give way if he is intent on mating with one of his cows, or on his way to attack a rival bull. Oblivious of snarls and bites from his other 'wives' he will rush forward, himself open-mouthed and hissing with emotion, beating his way past mother and child, and I have even seen him seize and fling aside a calf in his path. Or he will ride over it as if it did not exist.

Such behaviour was studied on the Green Holms of Orkney. On the north coast of the Muckle (Great) Green Holm seals haul out on a shelving beach of large round pebbles so convenient for a surf ride ashore without harm to their tender summer-fat bodies. This beach gives access to a shallow glen with a lochan and stream along which lie pools and creeping willow scrub. The breeding seals advance up this glen and fan out to their territories over the higher ground of this seventy-acre uninhabited island which is covered with thick soft grass.

The Kirkwall launch cast our party loose in a dinghy on the south side of the Muckle Green Holm one October morning, so that we might land unobserved by the concentration of inland-breeding seals on this north-sloping island. It was a warm sunlit day with a light south-westerly breeze, such a day as occasionally makes Orkney seem a halcyon land. The sea was almost still, deep blue, and dotted with 'skarves and dunters', as our Orcadian boatman called the black shags and the pied eider drakes and their brown ducks. We hauled the dinghy up on a tiny beach under low cliffs, and crept over the hill, through the long withered grass where starling and snipe rose before us.

Crawling forwards in waders and neutral-coloured windcheaters we established a strategic watch-post above the glen unobserved by the seals below. Some thousand grey seals lay scattered on both sides of the little glen bisected by a natural boggy drain-way.

Figure 5 is sketched from an enlarged photograph, and shows the approximate composition of part of the breeding herd. Most of the master bulls had a 'harem' of three to seven cows around them. A few big bulls lay alone, on the higher ground, as if either resting from the sexual struggle or unable to attract cows to join them.

The nearer scene was peaceful at first. There was small constant movement: one or two adults slouching down towards, or up valley from, the sea and the large lochan beside it, some 400 yards distant. The rich air was resonant with the musical complaint of calves, and the occasional moans of cows engaged in sundry domestic affairs calling for

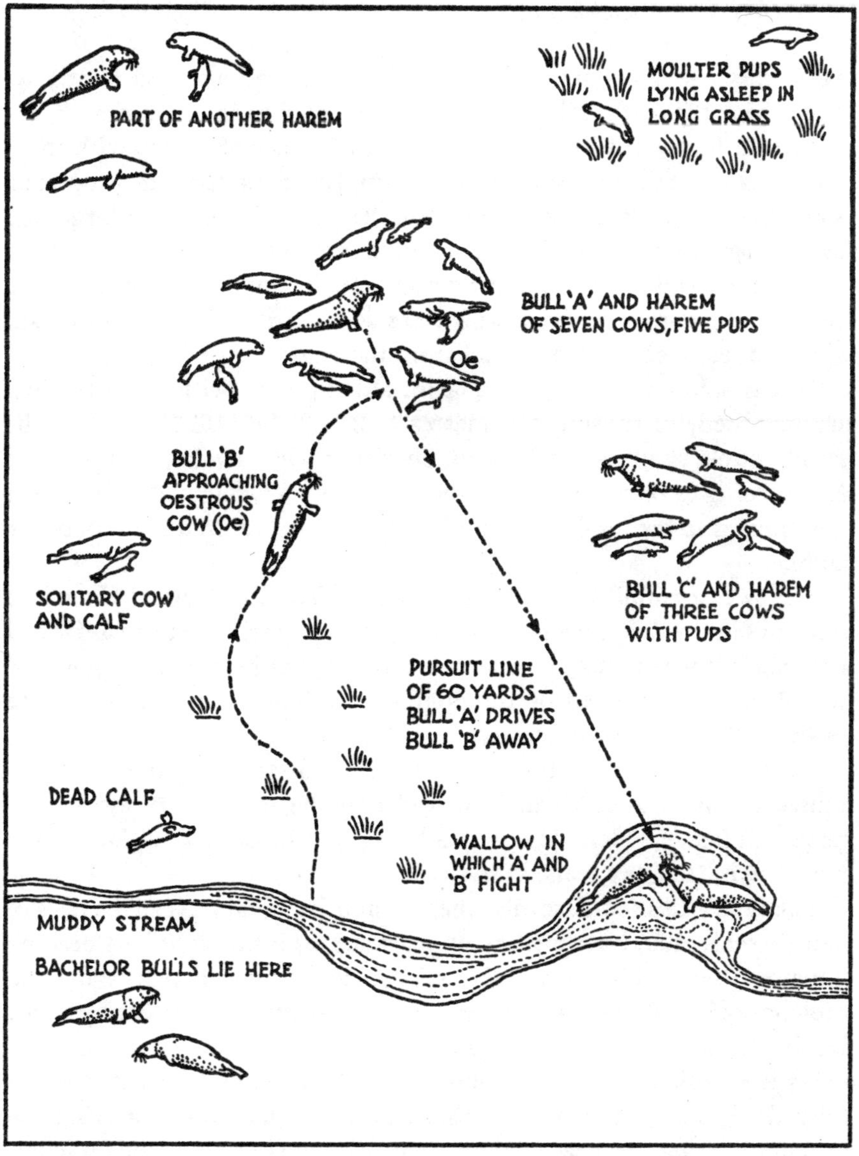

Figure 5 (not to scale). Part of a large inland grey seal rookery on Green Holm, Orkney.

vocal dissent. A skylark was twittering in the still air, and faintly we could hear the surf on the western cliffs exposed to the open sea.

Within fifty yards below us a huge bull (A) was sleeping, with seven cows and five calves grouped around him. Lower on the same slope and some twenty yards nearer us was another bull (B), equally large and moving slowly uphill, with many cautionary pauses, towards A's harem. We had arrived at an interesting moment. It looked very much like a raid by B upon A's wives, who were so placidly sleeping with their children, each cow a little apart from the next.

It was a cow who gave the alarm, although not with the intention, we surmised, of rousing the master bull. As the bull B reached the family circle she raised her head, suddenly wide awake, and moaned her warning to B to keep his distance from her calf. But evidently he was determined to try his fortune, and pressed on to within a few feet of the couple.

She swung round and flippered at him, hooting louder. He rolled sideways in the defensive position, silent but flippering in return. He did not snarl; it was almost as if he were afraid that his stealthy approach would be soon detected by the dominant bull and that silence for the present was the wisest procedure.

Two cows nearest B also raised their heads to study the situation, roused by the first cow's moans. And now the master bull awoke. He raised his head, which we saw was hoary or whitish, indication of ripe maturity in the grey seals, as in man.

The hoary-headed A sniffed the air, and his bleary eyes (the eyes of mature seals often have a filmy appearance, as if the sight had become impaired) seemed for a moment or two to be slow in registering the presence of bull B, which he could soon almost certainly detect by scent. But he did not delay much longer.

With surprising agility for his bulk he bounced forward directly towards B, who, as if only at that moment fully aware of A's close proximity, raised his head higher to stare for one second at the attacker. A had the double advantage of being in possession of the territory of the nursery, possession which always seems to confer a moral right and strength, and of being uphill of B.

As A did not pause for the usual formalities of static warning display of snarls and flippering, but came forward with faster bounces, crashing over the top of his cows and their calves, B turned and fled downhill. We expected A would, as is usually reported of master bulls, soon give up the pursuit once the intruder was clear of his territory.

Not a bit of it. The smooth grassy surface made the chase easy. The hoary master A vigorously chased B at least sixty yards, and in doing so skirted the harem of another bull (which we shall call C) containing three cows with calves. All seven members of this group stared at the chase with their huge brown eyes full of curiosity, and the nearest cow moved closer to her calf. B plunged down the slope into the boggy bottom of the glen, splashing through a much used muddy wallow. He had panicked, as well he might, for A was close on his tail; and as he tried to climb out of the wallow over the steep bank below the rising western slope, A caught and gripped him in the rear.

This was the limit. B was now above A, momentarily arrested by the steep bank and A's jaws. B turned and fell on top of A. They tore at each other with gaping jaws and loud snarls.

Both were soon plastered with the dark brown mud, and it was not easy to see how much damage each inflicted on the other. The intense mêlée lasted perhaps for a minute – which is long for a serious clinch, and time for many wounds to be given. Red blood flowed – much of it to be staunched by brown mud.

As if a boxing-match gong had been sounded for time, the bulls paused. They now faced each other, not squarely, but shoulder to shoulder with tails pointed in opposite directions. Each was in a position to bite the other's head and neck; but for a moment they were exhausted. They desisted and, rolling halfway over from the other, each bull raised the nearest flipper to parry any thrust of the jaws which would renew the battle.

This attitude is the typical defensive position assumed by fighting seals on land; by flippering and at the same time rolling sideways the attacked can evade the attacker's thrust. It is the seal's accepted boxing match stance.

My wrist-watch recorded that this battle in the mud continued for thirty-one minutes. There were at least twenty pauses or interludes of from 10 to 100 seconds between bouts of serious infighting. Yet when B at last broke away and retreated, not over the western bank (for he seemed too beaten and exhausted to climb upwards) but up along the rivulet, away from A's harem, he did not seem to be seriously wounded as far as we could see. The winner, muddy hoary-headed A, now made his way back briskly to his harem, passing close under our watch-post. He was bleeding from neck wounds, paused frequently, then bounced uphill with considerable vigour, a confident dominant male. He ignored the group under bull C, nor did the latter attempt to intercept A as he trundled past.

Twenty minutes later A was seen to copulate with one of his harem cows. He clasped her in his arms, his teeth gripping her neck. They lay sideways, reclining in the grass, his hind quarters pressed around hers. Her tail was erect. This union lasted twelve minutes before he relaxed his arms and drooped his head in sleep. She then turned towards him, mouthing gently at his head, as if anxious to mate again. But he was now completely passive except for one arm which, still half across her back, patted her each time she lunged affectionately; he seemed to invite her, by this slight flippering movement, to cease these now tiresome demonstrations of her desire.

The losing bull B lay a few yards upstream of the wallow. He was restless. Perhaps his muddy bloody wounds were troubling him? He raised his head, sniffing the air. There was one long slash on the right side of his thick neck, and blood oozed in a congealing blur close to his tail. He seemed a very frustrated male, yet still alert for another opportunity. However, he remained near the wallow for the present.

It seemed clear that in his attempt to invade the harem of the bull A, B had been attracted by the oestrous cow with whom A mated (perhaps not for the first time) after the fight; that B had scented the peculiar odour given off by a cow in season. The bull A had been asleep on the approach of B, perhaps after mating with this cow, since bulls, as we have seen, usually fall asleep for a while after and sometimes during coition. It is not proved that the condition of oestrus in a female seal is detectable by scent, but in many animals, from moths to domestic cattle, the readiness to mate is advertised over considerable distances by odours emanating from the female: and it is almost certain that this applies to the female seal.

It is convenient but not strictly correct to use the term 'harem' here. The bull grey seal however is not able to maintain a true harem, as the bull elephant and fur seals do; these round up their cows physically and herd them in harems before parturition, harems which can number up to 100 females at the height of the bull's fertility. The strength and potent behaviour of these huge males of the elephant and fur seals is prodigious. Without food for perhaps two months they are able to maintain physical and sexual dominance over this enormous harem, but these beachmaster bulls gradually grow thinner and weaker, and are eventually displaced by lusty unspent bulls.

The grey seal bull does not round up his cows. He merely establishes a territory where he maintains his authority only over those cows which enter it. All cows are welcome, but if, as they sometimes do, cows stray away across a spacious inland terrain, the resident bull does not follow them. A wandering cow may, Darling records, mate with a bull in a neighbouring territory, and perhaps remain there, perhaps return after mating to the first bull.

Our view of the thousand seals scattered inland over Muckle Green Holm confirmed this somewhat loose behaviour of the cows. Some cows were lying alone with their calf, away from any bull, or equidistant from two master bulls; they could, one imagines, be served by either, even by both in turn.

The behaviour of the master bull indicated only that he defended those cows within his territory with whom he was on intimate terms; and that when challenged by a rival bull poaching on his territory he would attack and pursue the invader for a sufficient distance out of that territory to ensure that further trespass was not immediately likely. He imposed his dominance by display and by fighting. But there was a limit to the distance he might travel out of his territory. In the case of A that limit was determined by the capture of B as he was trying to escape from the wallow. The muddy pool became that limit, where A thrashed B, and where B gave up the fight and retired, while A returned to base.

Territory on the wide open grassland is more elastic than on the narrow beach subject to tidal ebb and flow, and storms, and is more easily protected because the bull can travel easily and fast over the smooth grass, covering a distance which no bull would normally essay in pursuit of a rival over rough rock and pebble beach territory.

Bull A was evidently at the top of the hierarchy of seals breeding in that district of Muckle Green Holm plateau, and dominated more than his immediate nursery ground. This was demonstrated as he climbed past the little group consisting of the bull C lying with his three cows and calves. Bull C merely gazed (mildly or fearfully?) at A, as A skirted C's harem, but C made no attempt to intimidate or intercept the giant bull. By contrast C was a pale bull and looked smaller and more immature than A. Not only, therefore, did A dominate his own group of seven cows, but there was little doubt that he was unafraid of C, and could have invaded and possibly had his will with C's harem. But for the moment A was concerned to return to his own group, where one of his cows was in that condition of oestrus which demanded the exercise of his supreme function as potent male.

There were a number of bachelor or so-called reserve bulls lying around the outskirts of the main concentration of harems, some high up on the slopes, a few in the gaps between harems, and many in or near the sea off the landing beach. These were obviously waiting to fill any gaps left by master bulls spent in service earlier. They were most ready to challenge the masters when they detected a cow in oestrous condition. This is of course a natural insurance. If a master bull is too spent, has died or is otherwise unable to impregnate an oestrous cow, there would always be a reserve bull to ensure fertilisation. I watched one bull catch a cow in the surf here, and mate with her (Plate 4).

A similar pattern of mating behaviour is observed in the highly polygynous elephant and fur seal herds. The huge harems of the beach-master bulls are fringed with younger bulls in reserve, ready to copulate with oestrous cows overlooked by or unable to reach the lord of the harem. This bohemian behaviour is well described in *Studies of the Southern Elephant Seal* (Carrick, Csordas and Ingham, 1962):

'It is not clear how a beachmaster recognises a cow in oestrous, but this has to be learnt. Young inexperienced bachelors will attempt to mate with any seal smaller than themselves, not only with oestrous, lactating and pregnant cows but with weaned pups and immature seals coming ashore to moult in November, and even with corpses . . . the assaulted pup holds its hind flippers together and twists them away, biting at the bull [vide the description in the classic *Fanny Hill* of a virgin woman resisting sexual assault by a male roué by crossing her legs and biting him!]. Small groups of weaned pups are guarded by bachelors in the same way as beachmasters guard harems, and sometimes a pup is killed by a bachelor's bite.

'Even an experienced beachmaster sometimes tried to copulate with an unreceptive cow, especially after a fight early in the season, although it desists after a few seconds. One beachmaster, after chasing and pinning down a young bachelor, which was backing away, tried to turn it round into position for copulation. Relative size is the first clue used to recognise the opposite sex.

'All observers have remarked on how frequently breeding bulls in harems copulate during mid- and late October, when most cows are in oestrus. Six successful copulations in ninety minutes is not uncommon. A bull may copulate twice with the same cow, or one cow with different bulls.'

Certain long pebble strands under high cliffs or upon little-visited islands are selected by grey seal cows for their nurseries. Thus Abermawr at the western back of Ramsey Island in Wales, nearly 600 yards long, in some autumns has been the territory of five or six master bulls serving between them ten and twenty lactating cows at the height of the season, which is late September on this coast.

On Oronsay in the Inner Hebrides grey seals breed on beaches protected by weedy erosion platforms subject to submersion at high tide. In the deep water channels of these platforms leading to the beaches master bulls take up territory, policing the narrow waterway and checking on all seals using it. Intruding bulls are driven away, but cows are allowed passage to and from the sea, and are mated by the master of the channel when ready to copulate about fourteen days after parturition. This behaviour is an adaptation to the geographical situation.

The east side of Rusk Holm in the Orkneys has a steep beach about 500 yards long of large flat sandstone fragments; and on October 21st, 1963, we counted 147 cows, most of them with pups, and thirteen master bulls lying there, or in the sea hard by.

On the north beach of Muckle Green Holm, Orkney, composed of round pebbles, rock and sand, approximately 300 pups lay so thick in late October that they were, like the elephant and fur seal rookeries, two and three deep lineally from the sea to the top of the boulder beach above normal high spring tide (photograph, plate 7). The greatest confusion appears to reign when a boat passes close, or when men walk about in such a rookery. But there is a good deal of natural disturbance, too. There were many dead white pups on this Green Holm beach, as evidence of high mortality, due it would seem to overcrowding of nursery sites, as well as to the traffic of adults from the sea to the inland harems, which passed through this landing beach. Such a movement of the heavy adults frightened and dispersed and in some cases crushed to death new-born pups before they became strong and stout enough to resist; many lost their mothers in the confusion, and had doubtless died of starvation.

There is a beach of large pebbles footed at low water with small boulders on the west shore of South Ronaldsay in Orkney, looking across to the lonely one-croft island of Swona. It is not seventy yards long, unless one includes a southerly extension of rock slabs where a few calves are born in a sort of overflow nursery, an indifferent annexe. This is a delightful spot where rock-doves coo and clap from cliff to cliff, and come home to roost in a little cave which they have used for centuries. When the afternoon sun bathes the yellow rock strata and blue sea in glittering light it is very pleasant to lie warmly in the long grass of the low cliff and study this group. During October between twenty and forty pups were present. This beach nicely demonstrated the male seal hierarchy: it was the province of one master bull, one secondary or challenger bull, and one bachelor.

Visits over four days enabled us to sort out the order of dominance.

The beachmaster was distinguishable by his thick neck, his grey pate, and above all by his almost white nostrils. His patrol covered the best stretch of the beach, that is the area of thirty yards more or less free of large boulders where the cows could most easily come ashore, and where, therefore, the majority of their calves were deposited. He copulated with the cows in the sea along this stretch, and when not mating or sleeping he kept up his patrol by swimming close along the beach here. He did not come ashore. Many of his cows were asleep under water near him; their pale forms, couched in the cradle of dark seaweed, were clearly visible to us.

The challenger bull was a rich jet-black when wet, and brown-black when hauled ashore. He had lately received a severe wound by which he was easily identified: on the right side of his muzzle a piece of flesh sagged down, revealing a raw red gash about four inches long. His territory was the south centre of the beach where some half-submerged boulders gave him cover to retire in case of approach by the beachmaster. Here he would lie, often half-stranded by the tide. But he actively sought females at intervals, swimming in the open among the nurse cows which were inshore, either sleeping at or below the surface, or on passage between beach and sea, and some were suckling their calf.

In the midst of this busy scene he might pass unnoted by the master bull, on whom he kept a vigilant eye. When the master approached the south centre of the beach on his frequent patrol the challenger would disappear, usually fleeing out to sea, but sometimes skulking behind the inshore boulders. But on one occasion when the challenger was occupied with a cow who flippered at him coquettishly in the early stage of courtship, the beachmaster surfaced suddenly behind the couple, cutting off escape to sea. The challenger bull found himself grabbed in a torpedo-like attack by the master. There was an explosive fountain of foam as both bulls and the cow roared and flippered at each other. For a moment the master held to the challenger bull, who however twisted free and, his seaward line of retreat cut off, bounced ashore over the high rough boulders at an amazing speed – and probably some hard knocks and bruises. He was clearly terrified. He escaped over the rocks and well up the rough part of the beach, crashing over two sleeping white pups in his panic.

The beachmaster propped himself up on his fore-flippers, his body awash to his waist. His head swayed from side to side as he stared towards the other bull, as if assessing the situation. Then he slid back into the sea and made for the playful cow. To have pushed his attack on the challenger bull further the beachmaster would have had to climb laboriously over the huge uneven boulders. The geographical barrier was too much for him – and the challenger bull evidently knew this, and was able to rest in safety behind it. But the beachmaster had asserted his supremacy, and was content to return to his normal territory and activity in the water.

The challenger bull, we had noticed before, had frequently come to this spot where he now lay on the beach, which was easily accessible only at high tide direct from the sea, but at low tide, as now, protected by rough boulders.

Half an hour after the ignominious retreat to the beach of the challenger bull, a third bull, recognisable as a 'bachelor' by his much smaller size, with fewer neck wrinkles and almost straight cow-like nose, and by his overt behaviour, appeared from seawards, and swam slowly towards the scene of the recent encounter. The tide was flowing. He drifted towards a cow which was feeding her calf near the edge of the sea. This brought him to land within ten yards of the challenger bull who had, we thought, become sleepy as he reclined on the large pebbles, head pointed inland.

What would happen now? Would the secondary bull, recently defeated, retire before this new young male?

The challenger bull, possibly warned by the scent of the approach of the bachelor, suddenly swivelled round – after all, he had not been asleep, but only resting, alert, though with eyes closed. He stared at the other male, then roared and rushed agilely down the beach to attack the bachelor. We had not expected to witness such a nicely clear-cut demonstration of the escalation of power in bull seal society, but there it was. The bachelor scrambled swiftly back into the water and the challenger stopped at the edge, and in a few moments was reclining peacefully once more. We saw the bachelor frequently afterwards and always he swam or lurked on the fringe, ardent for an opportunity to reach an oestrous cow, but afraid of the established bulls.

I described a similar order of precedence of bulls on a Welsh beach hidden under high cliffs, in my book *Seals and the Curragh.* This beach – the Red Wilderness one, mentioned earlier – was about the same length as the South Ronaldsay beach, but held only half a dozen calves. It was dominated by a huge old scarred bull with a dark coat, Caesar; and a pale challenger bull, Shalimar, was confined to one end of the beach. Younger, bachelor bulls inhabited the outer rocks.

In that book I described the fighting which occurred when a large mature light-coloured bull, which had been present for a day or two at first, evidently hesitating to enter the master's territory, at last challenged Caesar, who was becoming spent. I called the lusty newcomer Brutus. The final battle took place in the water at the edge of the nursery beach, as Brutus sank his teeth into the folds of Caesar's neck:

'Spinning together in each other's arms the bulls seemed to be locked in an embrace of death – of death perhaps for Caesar, for Brutus was apparently not wounded at all. Blood mingled with the opaque wind-roughened water, as Brutus chewed at the open wound. The bulls sank abruptly, leaving the concavity of a whirlpool twisting steeply after.

'Caesar was first to shoot to the surface, gasping, in a mist of reddened foam. He was free, but Brutus was close behind. As soon as Brutus raised his head Caesar dived and attacked from below. Once more the surface exploded in foam as the great bodies revolved like dynamos. The struggle for supremacy had begun in the night and now a decision must be near. Once more they sank, leaving a grotesquely whirling veil of white bubbles at the surface. Brutus was first to breach this time. He was breathing wildly, his nostrils dilating rapidly as he sucked in air. His fine eyes were wide and white-ringed for a second before, revived with oxygen, he plunged like a bomb.

'I saw the ripple of his underwater attack move off in the direction of Caesar, who was floating exhausted farther offshore. This was Caesar's first sign of weakness – that he had permitted Brutus to come between him and the beach where his cows lay. The psychological effect of this re-alignment of positions was the turning point. The worn-out Caesar saw his enemy coming, watching the ripple gain speed; then suddenly he dived, and as he submerged I saw that his head was turned away.

'Two ripples moved in the direction of the open sea. Only one returned. Caesar was deposed. By the law of might he had given place to the fresh and ardent Brutus, who had waited many days for the right opportunity to attack and banish the physically spent master of the Pwll.'

Histological examination has confirmed these field observations that the older prime bulls, of upwards of ten years of age, are first to come into breeding condition in the autumn, and to establish territories. This is true of all the larger grey seal rookeries in the British Isles; and in the immense herds of fur seals on the Pribilof Islands off the Alaska coast, and in the elephant seal harems of the south. In the highly polygynous fur and elephant seals the old beachmasters come ashore first, and settle in familiar territories. Mature but less experienced challenger bulls arrive later, ready to mate with outlying cows on the fringe of the harem and to displace the beachmaster when he becomes too spent to drive the challenger off by threat display or force.

Last to arrive at the rookeries are the young bachelor bulls which assemble, generally in small groups, on outlying or non-nursery beaches or rocks. They have an obvious function: to mate with virgin cows which, without domestic cares, are observed at this season to frequent the same areas near occupied nurseries. Both young bull and cow, in meeting and mating near the rookery, are familiarising themselves with the breeding grounds for the future when the bulls will become challengers and the cows will pup for the first time. There are usually, however, in these gatherings of virgins of both sexes, more males than females, it is observed; and this applies to the assemblies of these age-groups I have counted in the grey seal. The explanation seems to be that the virgin, as soon as she is mated, may not remain in the assembly; but the young bulls stay on, still fertile and ardent to mate with other females. They form, with the challenger bulls, a reservoir of males to ensure that every female seal able to conceive is impregnated at her moment of oestrus.

A healthy cow seal, like many other female mammals, when bereaved of her infant at birth, or shortly after, and thus retaining her milk, comes into oestrus rapidly within a few days, and well in advance of her normal ovulation (in grey seals about fourteen days after parturition), and is mated thus early.

When the pup dies at birth, or is born dead, the cow grey seal usually ignores the corpse; it has shown no sign of life which will release fully the maternal affection. But if it has died after being suckled for a day or so, she, the grey seal mother may remain for a day or so, nudging and visiting the corpse. I have also twice recorded a cow tending a dead calf washing about in surf, but not at sea.

On cliff-bound coasts (Wales, Scotland, Ireland, Cornwall) where seals establish nurseries on narrow beaches washed by storm waves at high tide, new-born calves as well as older ones are frequently swept to sea and carried away. I have many times seen unweaned white calves floating far from the shore, with their mothers in attendance, but some have been alone. The lost little one must die of starvation, or by drowning, if it fails to swim back and find a beach near its birthplace where it can scramble ashore and moan until its dam, seeking for it at the nursery, hears and comes to it in time to feed it.

Cows which leave the nursery beach through such events may wander far either with the swimming calf or in search of it. In either case they pass out of the territory and surveillance of the beachmaster bull and are free, like the wandering oestrous virgin, to mate with any fertile unattached bull in the vicinity. Some of these tide-borne cows and calves strand themselves on beaches far from the birthplace, often on mainland beaches where they may be molested by man.

On September 20th, from a small boat, I watched a cow swimming with a white calf, attended by a young-looking smooth-necked bull, about a mile south of Ramsey Sound. They were moving south on the outfall of the tide, which here runs at seven knots. The bull was obviously keeping with the cow in order to mate with her – he had in fact established a moving territory around her, although no other seal challenged him except the mother, who flippered at him petulantly when he swam too close. He would dive when she beat him off, surfacing again a few yards away. She kept at the surface, nosing her calf constantly, as it floated in the calm current. It was in good fat condition, about a week old, and not in distress. It made shallow dives as I rowed the dinghy towards the trio. I followed them for three miles out into St Brides Bay; on this set of the tide the trio might drift in four hours south-west almost to Skomer Island, but at the turn of the stream they would be set drifting north and west into the open sea in the direction of Ireland. The cow was making no attempt to guide the drifting calf in any direction whatsoever, but seemed to be enjoying, as we were, the calmness of the autumn day at sea.

One wondered exceedingly how far mother and child together would drift into the Irish Sea, whether the week-old calf would survive the long immersion without resting on land, or the storms which subsequently were to follow this perfect day. I wondered, too, how long the bull would be willing to follow the cow out of his native sea, and whether, in about a week's time, in the excitement and heat of mating and the long sleep in coitus, the cow might drift with the bull far from the calf and lose it altogether. But might not the calf stay close to the mating pair out of instinctive fear of the open sea and its desire for maternal protection and food? So many ifs and buts. Such speculation is fascinating to the student of wild seals, and suggests fresh avenues of research and observation. Many of the pups marked in west Wales have been recovered in Ireland, some of them within a week or two of weaning; and it is a matter of speculation if some of these records have not been due to drifting on tide and wind-blown waves, accompanied part of the way by the mother, rather than a genuine migration. Most of the recoveries of these young pups are of individuals washed ashore dead or dying.

The young grey seal in its first few months of existence is proved to cross at least fifty miles of open sea between west Wales and Ireland, and 100 miles between Cornwall and Brittany. These open seas must be dangerous for the slow swimming pups; there is risk, not only from storms, but from sharks and killer whales. We discuss the predators of seals in a later chapter. While the adult seal might be agile and tough enough to evade sharks and killers, the little feeble-swimming calf would stand no chance. But in fact we know very little about the open sea enemies of the grey and common seals. Shark, killer whale and seal have survived in a naturally balanced community at sea, for the relation of the predator must be such as to ensure the survival of both; and so the numbers of predators must be controlled by their food supply.

8 The Common Seal at Home

As the common seal is less dependent on the land than the grey seal, it is correspondingly difficult to study, but where there are cliffs overlooking a nursery site their breeding habits can be observed. In Shetland there are several such sites, where the first detailed observations were made by such writers as Edmonston (1838), and Selous (1905) on the *Tangfish*, as the common seal is called in northern Scotland. 'Tang' is a Shetland word for seaweed, and this seal, as elsewhere in the North Sea, likes the shallow water where, under Shetland cliffs, the long strands of the oar-weeds cover the rocks, and act as a shelter, especially for the young pups.

I have watched this charming seal in Shetland and in Orkney in such rock-bound habitats, and in other diverse situations: on the smooth sandbanks of the Wash and of the Dutch and German Frisian Islands, including Heligoland; in the shelter of the Inner Hebrides; and along all four coasts of Ireland, where the terrain varies from open sea sands in the east to salmon estuaries and rocky embayed islands in the west.

The best account of the behaviour of the common seal I have read is that of L. S. V. and U. Venables, who, living in Shetland for a while, studied a colony near Fitful Head. Steep-to cliffs and deep water, and beaches of huge rough boulders are avoided by this seal, which cannot haul ashore in such situations. The Venables found that over 300 common seals would be hauled ashore in Noss Bay over low tide in calm weather in June and July; but much fewer, although always a few, remained hauled out over high tide, or when there was an onshore swell breaking on the site. With the common seals were often several grey seals but, say the Venables, 'the Atlantic seal is less prone than the common seal to lie ashore, and selects far more exposed sites for this purpose, so that the two species do not seriously overlap.' This is our experience in watching the two species in Irish and Scottish waters; although the common seal is better adapted for the marine life, and less dependent on the land for a nursery for its pup, it will come ashore at every ebb-tide if it can, and will give birth to its pup there under normal conditions, i.e. undisturbed by man or storm.

The birth of the pup and its early life over the lactation period has been described in Chapter 2. As soon as it is weaned it is in good coat, and hunted by Shetland crofters and fishermen, who can sell the skin to the trade locally, chiefly for making sealskin gloves, shoe coverings, etc. These pups are taken by the hunter as they lie out at low tide, and they are also shot as they float sleepily at the surface of the sea in shallow water, where the body can be retrieved with the aid of a long-handled hook.

But at this period the adult coat is beginning to fall out, and the hair and fur is loose in August. The old seals haul out in large numbers and are sluggish. However, the moult appears to be rapid, because common seals are recorded as mating as early as late July in the Wash; but the Venables describe mating behaviour only from September 5th onwards in Shetland, and this was all over before October 9th. This is the true copulation period, after the moult is completed; but the Venables also witnessed rolling amorous behaviour between couples in the spring (see page 58), resembling that of grey seals. It took place in April and May; in the grey seal spring courtship is seen from February to the end of April.

Both adult and sub-adult common seals take part, and this spring play is much more boisterous than in the grey, with much 'writhing and twisting in almost continuous contact, mouthing each other's napes, yelping and snarling and slapping the water with their flippers until the sea foams. Sometimes one member of a rolling pair will have the penis extruded'. A single seal will be stimulated to indulge in this behaviour alone 'and this often leads to a dual performance'. Full prolonged coition was observed but rarely by the Venables at these spring sexual games.

In June, as the pupping season opens, this interlude of wild courtship behaviour is over. It does not recommence until the moult is complete in September. Rolling behaviour is then resumed, and 'the blowing of bubbles introduced into the sequence. While rolling the pair coil downwards towards the bottom, either diving head-first or sinking tail-first, then each blows a fountain of bubbles and they float up together on the tail of the bubble-stream. Alternatively they remain on or near the bottom when, at least on some occasions, coition follows'.

Tame orphan common seals we have had, have blown bubbles under water when they were feeling playful, and it would seem to be part of the expression of *joie de vivre* in this seal. The Venables record that 'the chain of bubbles floating upwards and "boiling" at the surface is extremely conspicuous: possibly in the present behaviour sequence it is a form of self-advertisement. Similarly an excited seal makes its whereabouts obvious by bubble-blowing with lips awash creating a penetrating rasping noise, or causes rifle-shot reports by slapping the surface up to half-a-dozen times with its fore-flipper before diving'.

The moment of true pairing has been recorded by few observers; but it appears, as might be expected, to be almost identical with pairing in the grey seal, although less prolonged, and to take place almost invariably in the water, at or below the surface. The male rides the back of the female, clasping her under the arms with his clawed fore-flippers, and the union is effected, according to the Venables, by both twisting their 'hind-ends over sideways. Rough water was avoided by mating seals. No sexual behaviour was seen ashore'.

(On the other hand I have seen grey seals mating in heavy seas. On one occasion I watched a pair of greys firmly attached in coitus being washed up and down a rocky gulley, bumping against its rough cliff, apparently oblivious of the battering of their bodies by the heavy surf. Sexual behaviour ashore is characteristic of the grey, as we have seen in the chapter on their breeding.)

In all their observations on the rolling display and coition of the common seal the Venables saw no change of partners, such as occurs in the spring games of the grey seal. This would suggest that the bull and cow common seal are faithful to each other, at least for one season. But other writers consider that the common seal is polygamous, that a bull may serve several cows, or a cow may mate with more than one bull. As the common seal bull holds no land territory or nursery site, but lives in a community of all ages and both sexes throughout the year, haunting the edge of the land, and carrying out all sexual display and mating in the water, it is extremely difficult to follow the order of dominance in this highly social and aquatic species. Fighting appears to be rare. They are a happy people, and have only man as their principal enemy. His hand is everywhere against them, with often a price on their heads. If they multiply too much they are shot down ruthlessly. But still they survive – and man's predation may in the end be a healthy one, so far as it has gone today; his pressure has removed a surplus population, leaving abundant space and food for the better living of the survivors.

9 Population and Vital Statistics

How long do seals live? What is the proportion of young to mature seals? Are females longer-lived than males, as in the human species?

The common seals observed by the Venables in their admirable studies included a colony of approximately 400 adults and sub-adults at Fitful Head in Shetland, excluding pups not born at the time of the count, which was during the pre-pupping mating period of April and May, and therefore at the moment of lowest numbers in the annual cycle. In 1954 they estimated a minimum of 91 pups were born to these 400 older seals = 23 per cent, or allowing for some missed, one pup to four adults. If 400 adults (and sub-adults) produce 100 pups annually, and supposing the sex ratio to be equal, as is so far proved to be the case, then the population in spring, before pupping begins, is estimated to be:

100 pregnant cows
100 adult bulls
100 sub-adult females
100 sub-adult males.

I had independently arrived at the same ratio of one to four for the grey seal. In a paper I read to the International Union for the Protection of Nature, meeting in Copenhagen in 1954, I stated (*Oryx*, Volume 2, No. 6, page 385) that a minimum of 5,000 grey seal pups were probably born in the British Isles at that time, and that these were probably, in round figures, the product of:

5,000 mature breeding cows
5,000 adult bulls
10,000 immature and non-breeders.

The ratio of one pup to four older seals is, in fact, too high. A ratio of just over three proves to be an acceptable figure for estimating world populations of seals from known counts of pups: immediately after pupping one can reckon one pup to at least three older seals.

Studies of other seals have shown a similar ratio. In a study of the Alaska fur seal *Callorhinus ursinus* on the Pribilof Islands, the United States Fish and Wildlife Service has records of the herd of about one and a half million of these seals over fifty years of management and marking. About 530,000 pups are born annually; and some 60,000 large males are harvested each year. As the population is now stabilised, the annual total mortality (all ages) must be about 530,000 less 60,000 = 470,000, which is around 32 per cent. The mortality is naturally highest in the first year, and estimated to be 60 per cent.

Compare this with trends in the population dynamics of the southern elephant seal *Mirounga leonina* which has been studied at South Georgia and Macquarie Islands by marking individuals. Mortality in the first year is estimated at 40–50 per cent in both sexes. After the fourth year there is a higher mortality in males until they become powerful harem masters, when they may go on to live to be twenty years old or more. But the average breeding female is unlikely to live more than ten years; she is worn out much earlier than the grey seal. The bull elephant seal can weigh as much as 3½ tons and measure over twenty-one feet in length, but the female is only one-quarter this size, and suffers considerable rough treatment during the breeding operations at the harem of a beachmaster bull, even to occasional crushing of her skull during copulation.

From these two studies of large rookeries of seals, and from other longevity records and estimates, it has been possible to work out a provisional life table for the grey seal, based on a 60 per cent first-year mortality and a pregnancy rate of 80–90 per cent of mature cows, plus, of course, the supposition that the population is stable. Humphrey R.

Hewer, whose work in the field combined with histological examination, has advanced our knowledge of seal breeding physiology so much, has published such a table, and I am indebted to him for permission to reproduce it here (Tables 1 and 2).

Since he published these figures, Hewer has found a slightly higher pregnancy rate in cows he has examined, and this, combined with the fact that the grey seal at present is an expanding rather than a stable population, would alter the tables a little, although not significantly.

In both the southern elephant and the Alaska fur seal man's harvesting operations, directed chiefly against the larger males, has resulted in a somewhat earlier maturity in the young males and more space in the nurseries for females. Males obtain access to harems earlier because of the shortage of master bulls; this artificial situation affects the true figures for longevity and life tables. But while the average expectation of life is very low, probably less than eight years for the weaned pup, individual fur seals of both sexes are recorded (from dentition layers, or from marking of the skin with an identification number) to live at least twenty years.

Although the fur seal cow dies comparatively young, possibly worn out by half a dozen breeding seasons under the powerful beachmaster bulls, the grey seal cow is believed to live longer than the bull. Hewer found among 300 grey seals examined that the oldest cow was thirty-four years old when shot and the oldest bull twenty-three. In captivity the records are: a cow which died in her twenty-eighth year, and a bull in his forty-first year.

Comparable records for the common seal are scarce, but it is believed that this seal produces its first pup at three and certainly at four years of age, and that the bull is fertile at five years, possibly a year earlier. Age determination from sectioning of the canine teeth is successful in this species, too. In captivity a common seal has lived for nineteen years.

Reliable estimates of the age of seals can be made from sectioning the teeth, and counting the annual ridges or bands of dentine laid down. Due to uneven deposition (thickness) and quality during moulting, feeding and breeding periods, each phase can be recognised when a canine tooth is transversely sectioned from the tip of the pulp cavity to

TABLE I

Life-table for grey seal cows

Year	Number at beginning of year	Percentage mortality	Numbers and groups	
0	1,000	60	1,000 Pups→ Moulters	
1	400	30		
2	277	12	1,362 non-breeding cows	
3	244	6·7	(1,149 are immature)	
4	228	6·7		
5	213	6·7	Capable of conception	
6	199	6·7	Capable of bearing pups	
7	185	6·7		
8	173	6.7	868	
9	161	6.7		
10	150	6.7		
11	140	6·7		
12	131	6·7		
13	122	6.7	613	
14	114	6·7		
15	106	6·7		
16	99	6·7		2,500 cows
17	92	6·7		of breeding
18	86	6·7	432	age at
19	80	6·7		80 per cent
20	75	6·7		pregnancy
				give
21	70	6·7		2,000 pups
22	65	6·7		
23	61	6·7	306	
24	57	6·7		
25	53	6·7		
26	49	6·7		
27	46	6·7		
28	43	6·7	215	
29	40	6·7		
30	37	6·7		
31–34	66	?	66	

TABLE 2

Life-table for grey sea bulls

Year	Number at beginning of year	Percentage mortality	Numbers and groups
0	1,000	60	1,000 pups. Moulters.
1	400	30	1,362 immature bulls.
2	277	12	
3	244	6·7	
4	228	6·7	
5	213	6·7	
6	199	6·7	691 mature bulls most without territory.
7	185	10	
8	166	15	
9	141	25	
10	106	40	265 territorial bulls.
11	64	40	
12	38	40	
13	23	40	
14	14	40	
15	8	40	
16	5	40	
17	3	40	
18	2	40	
19	1	40	
20	1	40	
21	1	?	2 idle or senile bulls.
22	1	?	
23	0	?	
24	0	?	

Note. The assumption is implicit that *all* bulls above the age of ten years take part *annually* in the breeding colony.

the edge of the enamel. This is clearly shown in the large canine of the elephant seal (sectioned by Carrick and Ingham), in which these periods are measurable from the different densities and colours of the dentine, which is hardest at the prime feeding period, and poorest during fasting. Hewer has worked out a similar method of determining the age of grey seals by counting the annual rings in the cement deposited on the canine teeth; this can be done with some accuracy, and thus, by examining the stage of sexual maturity in the individuals from which the teeth were collected, the age at which breeding takes place has been ascertained. Hewer concludes that, as already shown in the Tables 1 and 2, cows are first normally able to conceive at five years; and bulls are fertile at six although they may not be powerful enough to hold a nursery territory until they are ten years old.

This late maturity in the grey seal contrasts with the precocity, already mentioned (page 59), in the elephant seal in South Georgia, in which Laws found that the females are sexually mature at two years of age; although on Macquarie Island the females of this same seal are not mated until their fourth or fifth year. Around both islands there is plenty of food in the sea, but at Macquarie the younger seals find it difficult to enter the breeding rookeries in competition with the dominant, experienced older cows and bulls. The overcrowded unexploited Macquarie population of the elephant seal is therefore 'self-limiting': the South Georgia group, exploited by man, with abundant nursery space, and shortage of beachmaster bulls (killed for blubber), is an example of a population adapted to optimum breeding success – and an example, provided man's control of its numbers continues to be within the seals' capacity to maintain themselves at the right level, of wise economic conservation benefiting both seal and man. This conservation aspect is discussed in the next chapter.

The (at present) unexploited Macquarie Island and other populations of elephant seals may well be responsible for the reported appearance or reappearance of this species in fresh localities, seeking new breeding grounds. Its world population in fact seems restricted at present by the limited number of suitable breeding beaches within its wide range over the cool circumpolar seas between thirty and sixty degrees south. In the words of Carrick and Ingham, who have studied this seal at Macquarie, this lack of adequate nursery ground there has resulted in 'a considerable deferment of breeding beyond the age of puberty'.

The human parallel in 'civilised' man is not dissimilar. In our overcrowded human society puberty is not normally the age at which male and female establish a breeding 'nursery'; there is usually a delay of from five to ten years, due to economic pressures, before man establishes a home-territory where child-bearing takes place. Precocity, or conception at or soon after puberty, only occurs where the population is low in relation to available territory and food supply, as with primitive tribes of man in thinly inhabited country (and as I found in studying rabbits placed in enclosures: breeding was early and prolific in the thinly inhabited enclosure, but was late and with high mortality in the overpopulated one).

Colonial breeding in animals offers a fascinating study; and biologists do not always agree on the interpretation of social behaviour. Is social breeding more successful because the individual is both protected and stimulated by the mass of the herd, which deters or drives away predators? On the other hand it is well known that predators assemble, even breed within or near a herd, because it provides a continuous food supply. The most successful breeders in a social group are invariably the older experienced adults which, highest in the hierarchy, occupy the best central positions, and are thus protected from peripheral dangers by the younger and weaker individuals flanking them.

There is F. Darling's theory that the social stimulation of the larger breeding group results in greater success, synchronisation and efficiency compared with the smaller group, and that there is a threshold of numbers in a social species below which breeding is unsuccessful. On the other hand there is the concept advanced by Wynne-Edwards, that colonial breeding is a mechanism which regulates the size of a population to a level consistent with the food supply: when the breeding group becomes too dense, the mortality rate increases, and the recruitment of new breeders is slowed down.

This seems to be true of the Farne Islands grey seals, which have been studied by Hickling and Coulson. Here colonial breeding under close protection has resulted in an increase in the herd from 100 in 1900 to 3,500 sixty years later. It is estimated that 40 per cent of the present mortality of the pups here is induced by overcrowding, and that even the health of the surviving pups is reduced by this pressure for territory. Social breeding in these 'slum' conditions is therefore a disadvantage. The puzzling feature of the sociality of the Farne grey seals is that they continue to crowd together to nurse their pups on certain of the islets only, leaving other islets, apparently equally suitable as nurseries, unoccupied. They may also move to new islets, abandoning one formerly occupied as a nursery.

Much the same can be said of other large British colonies. The mortality in the 2,500 grey seal pups at present born annually on North Rona is high, due to concentration of the nurseries on only 100 out of the 300 acres of grass, resulting in trampling and frightening of pups in the congested area of the access routes by the traffic of adults to and from the sea. And yet, where the pup is born on the perimeter, well clear of the congestion, it may sometimes wander or be forsaken by the cow, and be lost by her and die of starvation.

Of the three islands in Pembrokeshire where grey seals might find the safest sanctuary – Skomer, Ramsey and Grassholm, only the first two are used regularly as nursery sites, which are either in caves or upon open beaches, both storm-swept. But Grassholm, uninhabited and difficult of access, seems the safest from both man and storms, as there are good shelving rock platforms where the seals haul out above high tide, and these give access to gradual slopes of grass, ideal – one would have thought – for a safe nursery. But until recently there has been no record of a calf born there. Why, I do not know. Up to 150 seals may sometimes be counted on the tidal rocks at low water at Grassholm. On October 14th, 1963, we found a white calf about fourteen days old there – but it could have swum with its dam from Skomer or Ramsey, where, under protection, the grey seals continue to increase. They may yet inhabit Grassholm as a nursery, if, as seems possible, the other Pembrokeshire sites, which are far less safe from storm, become too congested.

There are many smaller colonies, in caves and hidden beaches, along the mainland of Cornwall, Ireland, and Scotland and the off-islets of the last two, where, by occupying the safer beaches of easier access, and by direct persecution, man has forced the grey seal to use these storm-swept sites normally inaccessible to man. Rearing success here depends not so much on the threshold of numbers as on fine weather in which the pup can live through the first fourteen days when it is comparatively helpless in the face of rough seas. Here one may find an isolated cow feeding a pup far from the nearest assembly, making, with the attendant bull, the lowest possible threshold of numbers – one pair.

Small groups of apparently non-breeding adult grey seals are present throughout the breeding season in Brittany, Cornwall, North Wales, the Isle of Man, where no pups are seen, or but rarely.

Could not the explanation of social breeding be simpler than the theorists would admit? We have seen that seals are intelligent creatures, and that they behave as gregariously as man does. May not it be that they *chose to live together*, and enjoy life more fully, as does man, by living sociably? I have watched an old bull seal, waking up on the Red Wilderness beach from a long sleep, suddenly discover that all his near companions, alongside whom he had first fallen asleep, had moved away to the sea. Almost in a panic he pounded along the beach to join a small group still resting on the pebbles some fifty yards away; satisfied with their close proximity at normal safety distance (about one yard) he moaned at them, waved a flipper to indicate that his intentions were conciliatory, and settled blissfully to complete his sleep.

Although there are (page 49) pioneer seals, youngsters who go forth alone on migration within a month or two of birth, and these may eventually colonise a new site, perhaps after being rejected at the old, crowded nurseries, it is obvious that the majority of seals (and men) prefer to live together in close colonies, seeing much of each other, and such aggregations form the bulk of seal and human populations. Such concentrations or conurbations of course cannot thrive without adequate food and territory being available, even if these have to be won by physical struggle and competition.

The two hypotheses are not incompatible. Darling's concept of the social stimulation of numbers is evidently valid for men and for seals;

but carried to the point of overcrowding it becomes self-defeating, and thus we return to Wynne-Edwards' theory that social breeding is a self-regulating mechanism. Man himself today might learn from the study of seal herds. Like these his agglomerations are in danger of destroying his environment for his children by overpopulation on certain breeding grounds. He must regulate his numbers intelligently, but not by war and disease (fighting and disease in seals), but by voluntary birth control. The wise use of his resources of food and living space at the same time would then provide him with the maximum opportunities for the pursuit of happiness for the maximum number of people. In the next chapter we discuss further the relations of seals and men, and man's history of the control of seals; this one should perhaps close with the warning that man's greatest need is to control his own expanding population, even to reduce it drastically, if he is to continue to share the earth with other interesting, beautiful, and often useful forms of life – including the seals.

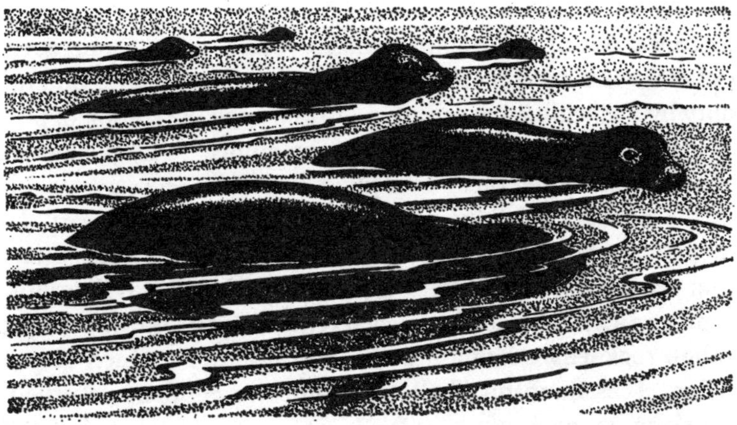

10 Seals and Men

Man has not ceased, even in this so-called enlightened age, to be the principal predator of seals. He kills large numbers deliberately: either harvesting them at certain rookeries with a rough care to preserve a breeding stock for the next harvest; or destroying them mercilessly as enemies of his own fishing interests, or for their fat or fur.

Here and there he is preserving them because of their low world population; and he has laws and regulations which provide a close season, or put a ceiling on numbers to be killed annually. But there are a very few sanctuaries where the seal is unmolested throughout the year.

Sometimes, out of his intense curiosity while visiting sanctuaries, as when photographing and counting them, man causes an unintentional mortality by disturbing a breeding group, causing separation of pup and mother, and a panic rush to the sea of the adults. In large colonies the adults may bunch together, trampling the pups severely, and in the case of the new-born fatally, as they bounce away to the sea. I have seen a group of twenty agitated adults, mostly nursing but some pregnant females, in an Orkney grey seal rookery, travel uphill in a dense mass, away from a visiting party, and crash down into the sea over rough rock ledges half a mile from the valley in which their pups were lying. It is doubtful if all these agitated mothers returned and found their pups again.

Human disturbance with consequent loss of pups through separation, and premature taking to the sea, occurs on the Farne Islands, North

Rona and other grey seal colonies, all nature reserves where the seals are protected all the year round; as well as on the annual visits to mark, study and perhaps weigh the pups. However, the scientific programme carried out at these sites over many years has produced much valuable information on the life-history and vital statistics of this species which fully justifies the temporary disturbance, and it is on such information that the future management of the herds can be based.

It is not necessary to describe here in detail the full history of man's great and greedy slaughter of fur seals for fur, and elephant seals for blubber oil, which in the last century reduced these animals to low populations all over the world; some of them are now beginning to recover, thanks to international treaties, and regulations for hunting them imposed by the proprietor nations. It is a story akin to that of whale-hunting – often carried out by the same hunters – of near extinction of some species. The pelagic whales and the pelagic seals (as distinct from land-breeding seals) belong to no one, and until the present international agreements (still unsatisfactory and often unobserved) limiting the numbers taken, were made, were on their way to annihilation. Some of the fin-whales may yet become extinct. In order to save them and try to restore other whales to sustainable numbers at which they may become once more an economic proposition for the now greatly declining whaling industry (which has cut its own throat in failing to moderate its appetite), it is now suggested that all whaling should come under the United Nations. Through this body all pelagic hunting of sea-mammals (which at present are common property outside the limit of terrritorial seas) should be controlled, and the first task would be to build up stocks. It is by no means an optimistic proposal, but highly practical; for it could eventually be a source of income to that impecunious body. Once a sustainable catch was built up the licensing fees for the taking of reasonable numbers could be adjusted to the cost of catching to the whale-hunting industry.

Seals which breed pelagically on floating icefields are threatened by fishers who hunt them today by boat, helicopter and light aircraft. Harp seals are killed in enormous numbers at pupping time as they drift on the ice between northern Norway and Canada; the stocks have been drastically reduced until today it has become essential to regulate their hunting by the three principal nations involved in this slaughter: Russia, Norway and Canada. Agreement now seems possible as I write this; but it does not apply to other ice-breeding seals (see page 150). All pelagic sealing therefore should come under international control; and the United Nations is again the obvious choice.

In the British Isles the grey seal was particularly vulnerable because of its dependence on the land as a nursery during the lactation period. There is a long history of the taking of this seal for its skin and oil, and for food. It was also, and still is, shot because of its alleged depradation on stocks of edible fish, and of damage caused to fishing nets.

Seal persecution is as old as the history of man. Bones of the grey and the common species have been found in middens and cave dwellings of primitive man along the coasts of northern Europe. As food and for its skin, used for clothing and leather (harness and cord), the common seal is still taken by Eskimo hunters in Greenland and Canada, as it has been from time immemorial. Yet, persecuted everywhere by man, the common seal has withstood his predation much more successfully than the grey seal.

In its main stronghold, the British Isles, the first protection for the grey species was slow in being applied. It began in the Farne Islands. Here, from the fourteenth to the sixteenth centuries, seals were taken for their oil and skins. The Durham Household Book, 1530–34, repeatedly mentions porpoises (*porco marino*) and 'seayle calffs' in its accounts. Pennant, some 300 years later, wrote that the tenants of the Farnes paid £16 per annum in rent, 'their produce is kelp [seaweed], some few feathers [of sea-birds, especially eider-down], and seals, which the tenant watches and shoots for the sake of the oil and skins'. In 1772 seventy-two calves were killed (probably, as Grace Hickling points out in her book on the Farnes, the produce of at least 250 older seals). Adult seals were also killed; one measured nine feet in length and weighed 658 lb, and produced twenty gallons of oil.

When Archdeacon Thorp purchased 'the Fern Islands' in 1861 he gave full protection to its seals and birds. His heirs founded the Farne Islands Association, and the islands today are the property of the

National Trust. Chiefly to protect the grey seals of the Farnes the first parliamentary act was passed – the Grey Seals Protection Act of 1914; this provided that any person who, between October 1st and November 15th 'kills, wounds or takes' or 'knowingly permits his boat to be used for the purpose of killing or wounding a grey seal' is liable to a fine not exceeding £10. But this period of protection of forty-six days covered only the beginning of the pupping period of the Farnes grey seal herd, which breeds much later than grey seals pupping elsewhere in the British Isles. In 1932 this anomaly was remedied by enlarging the period of protection to four months, from September 1st to December 31st. Ireland is excluded from this Act.

One hundred years of protection at the Farnes has seen a steady increase in the herd there from about 100 (1841 to 1920) to about 3,500 adult and sub-adult individuals today, a figure pleasing to the protectionists, but not so welcome to the salmon fishermen of the nearby Tweed and other north-east coast rivers, who shoot them freely.

There are numerous accounts of seal hunts in Scotland. From *Seals and Whales of the British Seas* (London, 1881) by Thomas Southwell we learn that from Orkney 'a ship commonly goes once a year to Soliskerry [Sule Skerry, thirty-five acres, about thirty miles off the north coast of Scotland] and seldom returns without 200 or 300 seals'. These were killed by knocking them on the head. From these dangerous visits in autumn, often over stormy seas (a forty-mile trip from Orkney) has arisen a good deal of the folk-lore and superstitions concerning seals, of which the ballad 'The Grey Selchie of Sule Skerrie' is most often quoted:

I am a man upon the land.
I am a Selchie on the sea.
And when I'm far frae every strand
My dwellin' is in Sule Skerrie.

(The full song used to be sung in Orkney within the last hundred years, and I give one version of it on page 171.)

The selchies (grey seals) of Sule Skerry have abandoned the islet since a lighthouse was built there; they appear to have concentrated on uninhabited North Rona, lying some sixty miles away to the westward.

Thomas Southwell also records that in North Ronaldsay, northernmost of the Orkneys, they take 'seals for the purpose of eating, and the inhabitants say they make good ham'.

Killing with sticks and stones was the usual method of taking seals at these battues. Martin Martin, in his *Description of the Western Islands of Scotland*, 1716, describes a seal slaughter on Cousamill, when all the routes of escape to the sea were blocked, and the 'passes surrounded and all (the seals) beaten to death with big stones'.

Captain Elwes, who visited the rocks of Haskeir west of North Uist on June 30th, 1868, informed Harvie-Brown that up to the year 1858 an annual battue was held there in the month of November, and that 40 to 100 old and young grey seals would be killed. After this annual hunt the Uist crofters believed that an old female *Ron Mor* (great seal) would appear in the villages and lament the loss of her relatives in the following Gaelic refrain:

> Ho-i-ho-o! Hoo-o-hao!
> Ho-i-ho-o! Hi-o-hao!
> Ho-i-ho-o! Hi-o-hao!
> Cha robh mise m'onar an raoir!

The first three lines of the refrain represent the moans of the bereaved seal; the last, literally translated, means 'I was not alone last night', and the several verses of the song itself, between the refrain, bemoan that she is the daughter of Hugh the son of Owen, a gentlewoman from another land. Obviously, from the name Hugh Owen, she was from the distinguished family of that name living in North and South Wales; and as, it happens, for many years I lived in the house of Orielton, near Pembroke, built by the Hugh Owen family originating in Anglesey, a haunt of grey seals.

An old woman of the Clan MacCodrum, North Uist, was seized with violent pains at the time of the annual seal-hunt on Haskeir, out of sympathy with her suffering relatives. My friend Seton Gordon writes of this legend: 'The surf-encircled rock of Haskeir or Hasker derives its name from the Norse, signifying the Deep-Sea Skerry. Here is the home of Ron Mor, the Atlantic Seal, that huge creature which in the folk-lore of the isles has human qualities. It is traditionally believed that these seals are the children of the king of Lochlann (Norway) beneath a spell. Their stepmother, jealous of their beauty, spent seven years and seven days perfecting herself in the Black Art. She then changed the children into seal form, so that they should be neither fish nor beast for ever, and that "their sea-longing should be for land, and their land-longing for sea, so long as wave beat on shore". But thrice each year, in order that their sorrow should remain undimmed, they were compelled for a brief space to regain their human form. It is said that a man of North Uist was one day walking along the western shore of that isle when, on reaching the entrance to a lonely rock-bound cove, he saw a number of seals casting aside their furry coats and bathing in human form. Unobserved, he picked up one of the discarded coverings and carried it to his home. The owner, too late, saw him make off with her coat. She was obliged to follow him, and when she reached his house he gave her human clothing and married her. Years afterwards, one of her children discovered the seal's coat, and brought it to her mother. She at once threw it about her, and entering the sea dived below the surface, after having sung a sad, strange song. It was said that from this union between man and seal-woman are descended the MacCodrums, of whom John MacCodrum, the ancient Gaelic bard, is perhaps the most renowned.'

There are other versions and elaborations of this legend to be found in the literature, and in *People of the Sea* (1954) David Thomson makes some pretty conversation with the Outer Hebridean people on the subject. Undoubtedly the legends are of Norse origin. The naturalist Millais observed that in order to propitiate Odin, the Norse god, at the time of the Haskeir battue, a horse-racing was held – 'a heathen tradition continued after ages of Christian culture, and after it has lost all meaning to those who observe it'. The MacCodrums were supposed to consume certain parts of the big seal, for reasons which Millais failed to discover; but when he was skinning a large grey seal a very old woman in a 'mutch' peered at him in the gloaming. She asked for the forepaws of the seal, saying her people had lived on seals for centuries. She proved to be of the clan MacCodrum. Another woman also asked for the paws, but would not have the hind paws. Neither of these women asked for the paws of a common seal Millais shot, but came specially to ask for the forepaws of a half-grown grey seal which he had killed on Haskeir.

In 1858 the brutal annual battue on Haskeir was stopped by Sir J. Orde. But there are records of later battues there: twenty-two killed in 1897, when a boat from Lewis also failed to reach Haskeir and was driven to Skye; fifteen were taken in 1898. According to present-day reports, although seals are still hunted in the open season in the Hebrides, few visits are made to Haskeir now, despite which the colony there does not increase – the grey seals seem to have moved to the more accessible Shillay, in the Sound of Harris.

Pinkerton, in his *Collection of Voyages and Travels* (1808–14), says that the natives of Harris salt seal meat in the ashes of burnt seaweed. This meat was eaten in the lean days of spring, using a long pointed stick instead of a fork, to avoid the smell which their hands would have afterwards. 'The flesh is pectoral [i.e. good for chest disorders], astringent and a remedy against diarrhea and dysenterea; the liver dried and pulverised is good against fluxes; the meat is even eaten by persons of distinction disguised as ham. In the islands southward [i.e. Uist] it is used as fish in Lent.'

I found that in the west of Ireland, where seals are still killed freely, the oil is collected from boiled seal meat as it drips from the suspended sack in a warm room, and is kept in a bottle, to be used as a cure for rheumatic disorders. Near Brandon Creek in Kerry I was offered a drop or two to cure a sprained knee.

Sea monsters in Shetland are for the most part called Finns, taking the shape of any marine animal and also of human beings. They may pursue boats at sea, and it was dangerous to say anything against them – the old seaman's dread of mentioning things, cherished or feared, by the right name at sea, which I encountered when on a trip to Rockall in a Fleetwood trawler. In seal form the Finns came ashore in Shetland to dance on the sands almost every winter night. They would cast off their skins, and act just like men and women. Some of these Finn women are said to have been captured and married by Shetlanders; and some folk in Shetland like to trace their descent from these mermaids and mermen. The old superstitions linger, and while the modern Shetlander hunts seals with a gun and a motor-boat, he is often careful to avoid mentioning names at sea other than 'him' and 'it' or 'they' in referring to the quarry.

In Ireland the clan of Coneely is said to be lineally descended from a seal-woman. At Ballyvaughan I was told that after one man had killed a seal his pregnant wife bore him a son with webs on his ears (one version was of webbed feet and fingers); after this no one would kill the seals again, and they have become numerous and tame, lying out with impunity even inside the harbour (as I saw for myself). When a 'foreigner' (that is, a man from elsewhere in Ireland) came fishing and was short of bait, and killed and cut up a seal for his lobster pots along the Ballyvaughan limestone cliffs, he found next day that his pots were chock full of heavy stones. As he hauled them the pots broke with the weight of the stones, and, terrified by these omens, he fled from the district altogether.

Paddy O'Connell of Liscannor told me how he shot, and thought he had killed, a big seal one day which had been eating the salmon from his nets as boldly as you could wish. The seal would be waiting at one end of the net for a salmon to get caught, while Paddy waited at the other end. Having fired the shot Paddy said to himself: 'And that's the end of you, me bhoy, and I'll have me own salmon in peace from now on.' But next day, going out to his net without his gun, Paddy saw the seal calmly waiting at the other end of the net, 'and would you believe ut, the creathur' has his arm in a sling!' Paddy's brother-in-law Gus confirmed this story that the seal had a bandaged arm the day after Paddy shot at it, and that when he told the news to the priest, 'Man, he just prayed the seal did not be badly hurt at all. For a clever seal like him, knowing when Paddy had the gun or did not have it in his boat, deserved a long, long life.'

Such tales were passed between us at the Anchor Inn, Liscannor. Farther west, the Wallis family of Inishturk, near Clifden, are noted fishermen, and told me where to find baby seals. High Island was reputed to be the home of a huge man-like creature of the sea, in fact a merman. One day, said Pat Wallis, he and some others had come ashore from fishing to stretch their legs on this hump of an island. They lit a fire to boil a kettle of tea, and he strolled up and over the high hill to look out west for signs of weather. It was very calm. Suddenly he noticed a huge figure 'twice as stout as a big man and twice as tall, and him resting with his tail in the still water, almost upright like a statue'. The girth of this creature was 'ribbed with rolls of fur and flesh, I couldn't tell which, but the creathur' was dark, like a blackamoor, and he had the face of a man one hundred years old, and he had never a whisker on him – like a seal has. He was a Roan-man, a merman you would say'. From the top of the cliff he threw some fine pebbles towards it, waking but not alarming it. It looked around, bewildered, but soon dozed off again. At last he hit it with a small stone. 'Then the merman plunged head first into the still sea, just like a human diver, and I never saw it again.'

When I retailed this story to a Coneely I met farther north up the coast of Galway, he asked what day of the week it was that the Wallis men saw the whiskerless merman on High Island. 'For in my opinion,' said Coneely, 'it must have been a Sunday, the merman had shaved hisself for the High Mass they do be having at the back of the wild Inish rocks of a fine Sunday, to give praise for being spared from the likes of fishermen who throw stones at them and do kill them to death with clubs and rocks. Myself I wouldn't touch the wild sea maids and men, and there do be plenty along this western shore.'

But then he was a Coneely. It would be unlucky for a Coneely to molest a seal.

A Courtmacsherry fisherman at Coolbaun Inlet also thought it unlucky to kill a seal; for, said he, 'they never touch a human body, but if they find one in the sea, they stand watch beside it, until someone comes'.

A fisherman from Schull, who was bending a new trawl to the warps on the quay at Kinsale, told me he caught seals now and then in his trawl, chiefly small ones, but because he knew it was unlucky to kill them he always let them go, but 'not before giving them a big whang, just to warn them from poaching any more'.

A man from Long Island, coming ashore to Schull on a sunny calm Sunday for the usual Mass, to be followed by a few jars of porter and talk from one inn to another, said to me that he did not hold with seal killing. 'Half-human they do be, the way they do be crying at the back of the island, on the rocks of Illaunricmonia. I heard them talking just now, as I was leaving the door of my cousin's house on the Long Island. It's a fine still day for the saying of Mass.' And he added quaintly, 'The *Roan* do approve of sunshine, for lying in it.' (*Roan* is the Irish or Goidelic for the great seal, as *Ron* is the Scots Gaelic for the grey seal.) So I walked the long road west, opposite the Long Island; with the tide low I presently could see the enchanting islands and islets of uninhabited Illaunricmonia lying in the sun, and some twenty seals were together there, singing and sunning themselves.

In connection with the habit of throwing stones at seals, there is a curious tradition, mentioned in Monk Gibbon's *Seals*, that doing so will stop the music of the fairies which are believed to sing to the seals. In *The Mermaid of Inish-Uig* (Tory Island) R. W. K. Edwards tells the 'true' tale of a poor woman who went down into a seal's cave and gave her child to the seals, she herself being shot in mistake for a seal by one of the local hunters as a result.

A farmer at Dunmore East told me he killed seals for sport, although he also collected the 25*s* subsidy for the snout. 'Ye must,' he said, 'kill the seal with a ton-weight blow on the snout, not the head, and they will die instantly. A strange thing happened to my neighbour one day. He killed a cow seal heavy with calf in the Seal Hole, and afterwards went back home to get some help to skin her. When they got back to the cave they found the cow had given birth to her pup, and, believe it nor not, both the snout from the cow and the calf had been cut off clean.'

Stories of strange seal-hunts, of supernatural events with mermaids and mermen connected with seals, are also to be found in the folk-lore of Cornwall, Wales, and the Isle of Man; in fact wherever seals are found. Their human-like faces and forms have given rise naturally to these. Despite this human appearance, seals were and still are shot, not reasonably for their skins and oil, but often for 'sport'. About 1889 Lord David Kennedy went shooting seals in Pembrokeshire for the fun of it. He shot seven, quite a sizeable bag, since they were not common on this coast at that time; but 'he never saw a male exceeding eight feet'.

In *The Home of a Naturalist*, published about 100 years ago, Edmonston and Saxby pronounce seal-hunting in Shetland a splendid sport. The hunter in Shetland provides himself with a 'waterglass' (a box with a pane of glass for its bottom), a 'clam' (enormous forceps with jaws two to three feet wide when open), and a twelve to twenty feet gaff with a ling hook lashed to it. Two stout lines are attached – one for lowering the clam open, the other for closing the blades over a dead submerged seal. The gaff is used in shallow water or when the clam has raised the seal near enough to the surface. 'My uncle the Laird of Buness shot 300 seals of mixed species in ten years.' Dogs were taught to retrieve shot seals before they sank; collies were the best breed for this task.

The lonely island of North Rona (300 acres) and the rocky islet of Sulisgeir lie near each other some forty miles north-west of Cape Wrath. After Rona was abandoned finally by its inhabitants in 1844, it was grazed by semi-wild sheep, which in the summer were tended by the Lewis crofters who owned them, during rare visits to collect the produce of the flock. After 1844 no late autumn visits are recorded until 1883, when it was found that grey seals were breeding on the island, and 107 were promptly killed there, although there is no record of a colony earlier. At this time, as from time out of knowledge, the men of Ness made an annual trip in a large sailing boat, which they pulled ashore, to Sulisgeir, in order to take the young gannets or gugas at fledging time in late August and September; and there is a record of thirty grey seals killed on the rocks of this inhospitable islet in 1882.

Perhaps it was the knowledge of this new seal colony that was partly responsible for a tragedy. In May 1884 two Lewis men sailed for North Rona to go into voluntary exile because, it is said, of a dispute over the appointment of a layman as preacher to the church at Ness. Despite pleas from their friends, they decided to spend the winter there to atone for their action against the minister of Ness. Ostensibly their reason for going was to look after the sheep, and to kill seals, according to Harvie-

Brown, who was to visit the island in 1885. Meanwhile no word was received of the welfare of the exiles, since no boat could land there before April 22nd, 1885, although two attempts had been made earlier by the men from Ness. When at last this relief boat arrived, 'No one met them. At the door of the little half-underground house occupied by the two men the boatmen found the body of Malcolm MacDonald in a sitting position beside an improvised fireplace, as if he had fallen asleep. On the floor of the house, beside the fireplace, lay the body of Murdoch Mackay. His tartan plaid was placed neatly and carefully over and under him, showing that the deft hands and warm heart of Malcolm MacDonald had performed the last sad office to the body of his dead friend.' A post-mortem revealed that they had died naturally, Mackay of acute inflammation of the lungs and kidney, MacDonald of cold, exposure and exhaustion. Neither man could write, and the only message they left was in the form of an ingenious calendar, which indicated the approximate date of MacDonald's death: this took the form of a bar of red pine wood, smoothly dressed, in which was cut a notch for each day, a deeper notch for Sunday, and a cut from side to side of the bar for the end of the month. 'The plan is simple, clever and intelligible. The markings begin on Friday the 21st June, 1884, and cease on 17th February, 1885. Towards the end the notches are less neatly and accurately made, indicating very clearly that the deft fingers which fashioned the rest were becoming weak and powerless to cut the hard pine wood.'

No record exists of how many seals were killed by the exiles in that autumn of 1884. No one has lived a whole winter there since, and the seals have lonely North Rona to themselves today, apart from occasional visits from a seal-observation party.

The story of the grey seals of St Kilda is not dissimilar. The human population was evacuated in 1930. In the literature of the island (as far back as 1697) there is no record of seals ever breeding there; the islanders exploited all natural resources, and killed the odd seal at sight Seventeen years later, on September 29th, 1947, during a flight in a Sunderland flying boat for the purpose of counting seals by aerial survey along the coast of Great Britain, with eight other observers, I flew over and circled St Kilda. We recorded: 'None seen hauled out on Village Bay shore, but 20–25 animals swimming about in Village Bay were probably seals. The grey seal does not breed at St Kilda.' In the winter of 1957–58 R. P. Saundby was stationed with the RAF Unit newly-established there, and counted five seal pups in Village Bay, thirty on Dun, and two in Glen Bay; and up to 150 adults. Pupping was from October 6th to mid-November, as on North Rona. The mortality of these new-born calves was, and is, very high, due to the severe Atlantic storm-waves which periodically sweep the rock ledges and boulder beaches on which these babies are born. As yet the pregnant seals have not reached the stage, as at Rona, of climbing up on to the grass for safety, but they may yet do so, as St Kilda is now a nature reserve where no animal is exploited by man. It took the North Rona seals about forty years of peace and quiet to come right ashore to breed.

In Ireland seals have never been protected by law, in fact in most areas where they appear they have a price on their heads, especially in Salmon Fishery Districts, varying from 10*s* to £3 per snout. Yet both species manage to maintain their numbers in the form of small groups along all coasts, perhaps assisted by reinforcements of young seals born in Wales and Scotland (see page 155).

As for the common seal, no protection has ever been afforded this lively intelligent creature, nor does it seem to need protection urgently at present. Yet one feels that, if only on humane grounds and to check much irresponsible amateur hunting, it ought to be given protection over the month of June, when the main fall of pups takes place. As we have seen these are helpless and can be taken very easily during the first day or two of their lives, as they lie out at low tide. This habit of pupping at midsummer is an advantage for survival, since the young seal has the finest weather ahead in which to be nursed and become independent. At the same time, with no close season, it can then be easily killed by men ranging the sandbanks and breeding places in the fine weather; and this is what happens. Many cows are shot then, pregnant or nursing; the pups which escape, drift about and die of starvation, and are a pitiable sight. I have seen lost pups 'sucking' at the belly of a boat in their hunger; and rescued and fed more than one. Others come ashore, often at a holiday resort beach; and the RSPCA has, as already mentioned, established *crèches* along the East Anglian shore where these waifs, if strong enough, are tube-fed on milk and whale oil until they are judged fit for release at sea.

For the rest of the year the common seal is difficult to shoot. Once it has been shot at it becomes extremely gun-shy for a long period. And in the British Isles it cannot be harvested in sufficient quantity to be of commercial importance, except very locally, as in Shetland. Occasionally an organised raid is made by a fur trader on a common seal assembly, as happened in the Wash in 1964; but apparently the result in marketable pelts, and the nuisance created by skinned carcases which were left and washed ashore, did not make a repeat attempt next year worth while. The surviving seals were said to have become wise to the appearance of the raider's boat, and difficult to get near after the first few days of shooting, even with guns fitted with silencers and telescopic sights.

Both species of seals are shot at, some killed, others scared away, and a few caught in salmon nets, at the mouths of salmon rivers in the British Isles. From time to time they return, even swimming far up some rivers, and there is little likelihood of their being even locally exterminated for good. Fishery Boards can obtain a local suspension of the Grey Seals Protection Act in order to shoot grey seals in rivers between September 1st and the end of the year; but in fact few applications for suspension orders are received, as there is nothing to stop marksmen from shooting at grey seals in the belief or under the pretence that they are killing the unprotected common seal.

Describing the behaviour of a seal at a salmon net, Inspector Francis Ryan of Carrick-on-Suir, Eire, said to me, 'When a seal appears he is always at the far end of the net, opposite to you. As soon as a salmon strikes the net the seal dashes in and tears at the fish, leaving only slime and scales behind. Once, when shot at, the seal threw the salmon high into the air, and dived after it.' And he added an observation I had heard several times before: 'You can always tell when a seal has killed a salmon – by the ring of oil which comes to the surface afterwards.'

There is no doubt, from the evidence of stomach contents, and from captures in salmon nets, that both species of seal take salmon freely, but by no means exclusively. Perhaps the most serious complaint comes from salmon fishermen who allege that a large percentage (but actual counts show on an average only 6 per cent) of the salmon caught bear the marks of seal teeth and claws. How one distinguishes marks of seal attacks from those of other predators such as otter, shark and other fish is not disclosed. In addition it is claimed that at least 10 per cent of the salmon nets are damaged by seals, resulting in loss of catching efficiency. On the other hand it is significant that it is in the years of greatest numbers of salmon landed by sea and river fishermen that the complaints against the seals have been greatest in volume. Obviously the seals have also enjoyed a good run of salmon and had a successful season – while running the risk of being shot by estuarine marksmen employed by the prosperous salmon fishery interests.

It must be intensely annoying for the fisherman to find a seal helping itself from his line and net, and this has often been recorded. At a conference held by the Ministry of Agriculture and Fisheries in 1935 which I attended one speaker reported that when he was fishing for pollack seals took each fish off his line as he caught it. I have often had the opposite experience: of seeing a grey seal close to my boat when I have been catching pollack and mackerel (and this happened to me again last evening when I was catching fish for an orphan common seal I am rearing), and both seal and I have enjoyed good hauls at the same time. Or, to be strictly accurate, the seal kept diving within twenty yards of the boat and I presumed it was fishing, as I was, on the edge of the Jack Sound current where the fish lie, although I saw no fish in its mouth when it surfaced. As I pulled in fish after fish, I whistled to the seal and hoped it was having a good meal – under the water. Certainly I have never had a seal take a fish off my line in these or any other circumstances, to my knowledge. Fishermen will say it is no use fishing where a seal is seen; but it is far more likely that the seal has come to the spot because of a good run of fish, and the human fisherman would be wise to be guided by the seal as to where to ply his rod and line.

How much edible fish does a seal devour in the course of the year?

Stevens made a rough estimate of 15 lb a day for an adult grey seal, which is rather less than 2½ tons in a year. But grey seals normally do not eat every day of the year. When breeding and when moulting they fast for days, possibly weeks. A 600 lb male Californian sea-lion in New York Zoo was kept in good condition on 14 lb of fish a day; and the ration for an adult grey seal is reckoned to be about sixteen herrings a day, which is 8 to 12 lb according to size. Estimation of the annual consumption of fish by seals is a difficult business, liable to extremes of error, so little do we know of this intake and its variability. For what it is worth Scheffer estimates that a healthy seal requires to eat an average one-fifteenth of its weight each day.

Seals are difficult to trap when they attack fishing-nets. Small seals are often drowned in nets; but large seals tear their way out, and thus destroy valuable gear. In Pembrokeshire the inshore fishermen claim that seals are now so numerous that it simply does not pay to put out a net overnight. In the mouth of the Shannon river in Ireland I was shown a drop door trap invented by J. Meehan. It was twenty feet by six feet, and fourteen feet high; at the closed end a fish bait was attached to wires. On attacking the bait inside the trap, the seal released the catch to the open door which dropped down behind it. Mr Meehan chuckled when he said that his home-made trap was far more successful than the copy made by the Irish Fisheries men! In the same area J. Daly, Clerk of Fisheries, told me that the fishermen had learned to exclude seal from their fixed stake salmon nets by reducing the entrance to the traps to three inches with steel bars, through which prowling seals would or could not force a way. The use of synthetic fibre (nylon) has greatly strengthened modern fishing nets – seals cannot bite through them.

A new objection to the presence of seals at fisheries has arisen from the discovery that both grey and common, together with probably all other northern seals which feed on cod, act as primary hosts for the cod-worm, the larva of which is ingested by the seal when it eats an infected fish. The flesh of cod, smelts, eels and in fact most species of ground fish, becomes infested with small nematode worms. The scientific name of the worm parasitising the cod is *Porrocaecum decipiens*. The adult egg-laying worm lives in the intestinal tract of the seal. The stomachs of adult seals often carry a considerable ball of live worms. The worm eggs are voided in the seal faeces, and the larvae, on hatching, swim to an intermediate host, such as the crustacean *Mysis*. This is eaten by the cod, and the larval worms work and feed their way into the muscle flesh of the fish – to be liberated and lay their eggs in the stomach of the seal which eats the cod flesh.

As the primary host, and largest and least numerous in the chain, the seal is the easiest link for man to break in the life cycle of the parasite. How to do this has occupied the attention of the Fisheries Research Board of Canada particularly, because the candling process at the factory by which the worms are detected in the fish fillets being prepared for human consumption was not satisfactory, leading to complaints by the consumer. (It is poor consolation for the consumer to be told that, provided the fillets are thoroughly cooked, the risk to health from ingesting the worms is nil!)

At first the grey seal, which has a population of only about 3,000 individuals in Canadian waters, was thought to be the main vector: its large stomach contains larger numbers of worms than the small stomach of the common seal, but as the pelagic harp seal is by far the most numerous seal in the area and also becomes infested with cod-worms during its stay in the Gulf of St Lawrence from March to May it may be that the harp seal will prove to be the most important vector in the end. The harp seal is already vigorously over-hunted in Canadian waters for its valuable skin: the value of pelts landed in 1963 was about £1¼ million from over a million of these seals counted in the Gulf. Short of extirpating all the seals of all species in the Gulf – and even this might not be effective, since the cod-worm may have alternative hosts not yet discovered – there seems to be no other solution to the cod-worm problem except greater efficiency in detecting and removing it from the processed fish fillets. Worms exist in the flesh of most fishes.

Apparently cod-worms infest all cod taken in British inshore waters, and the incidence is highest in cod caught in the vicinity of the grey seal colonies in the Farne Islands (32 per cent of marketable cod), Inner Hebrides (up to 85 per cent), Shetland and Orkney (up to 65 per cent). From the fact that cod taken thirty miles or more offshore and in deeper water where seals are rare have low or no infestation, it does look as if the cod-worm flourishes particularly in the vicinity of seal haunts and rookeries. Fifty years ago, when seals were much scarcer, worm infestation of cod was unknown in British waters, so its increase has been dramatic, although it cannot properly be related to the increase in the seals which has also undoubtedly taken place, until we have much more convincing proof. Inshore waters are now so heavily polluted by man with various effluents, including sewage and agricultural herbicides, that it is possible the cod-worm is deriving a biological advantage from a new source in this direction.

The case for a reduction of the numbers of the grey and common seals in the British Isles has often and forcibly been put by the fisheries' interests. A great number of both species have been shot, killed, wounded or driven away in the estuaries of the salmon rivers, especially in north-east England, Scotland and Ireland. Even explosives have been used to frighten seals. One would have supposed that these perfectly legal control measures would have been sufficient to protect these interests, but the outcry goes on – chiefly from fishermen along the coast from Tweed to Tay, and including the net fishermen who intercept the salmon before they enter the estuaries. And because of the success of protection in the nearby Farne Islands this outcry is aimed against the increasing herd of grey seals there.

In 1955 the Nature Conservancy was asked by the Ministry of Agriculture, Fisheries and Food to advise on this alleged damage; and in 1959 a Consultative Committee on Grey Seals and Fisheries was set up, on which both bodies, and several independent experts, were represented. It was decided to make an experimental cull of the Farne Islands herd during the breeding season, for which purpose it was necessary to apply to Parliament for an Order to take grey seals in the close season. It was agreed that damage to local fisheries might be reduced if over several years eighty-five adult cows with their calves were killed annually, or failing that, 360 female calves, or a combination of the two; this cull was calculated to reduce the breeding potential of the Farnes seals by 25 per cent by the end of the fourth or fifth year, since results could not be expected from culling female calves until that period had elapsed during which these would have become pregnant. Culling began in the autumn of 1963.

Licences were also issued from 1962 onwards for the taking in the breeding season of a limited number (originally 750 pups) of grey seals in the Orkneys, a traditional hunting place, by local men who were accustomed to selling sealskins to dealers. The price of sealskins had been rising appreciably at this moment. One hunter even bought a seal-rookery island for the purpose of farming its large herd of grey seals. I visited this island in October 1963. After seeing how severely disturbed the cows were here and elsewhere in Orkney at the killing of their pups in the white coat stage, I proposed to the authorities that licences to take pups in the close season in future should be restricted to the taking of the deserted moulted pups above one month old. In 1964 Orkney seal-killing licences were wisely restricted to the taking of pups in this condition of first adult coat. In two years already the considerable disturbance by hunters killing suckling white-coated babies had resulted in large scale desertion of some breeding grounds by bereaved cows.

None of the slaughter in the Farnes, or Orkneys, was accomplished without a tremendous ventilation of views in the popular newspapers, on radio, even on television, as well as in some scientific journals. Salmon fishermen approved, but the loudest body of public opinion, comprising the humanitarian, and many naturalist, organisations, nature lovers and ordinary people, were against the culling. In an article on 'Grey Seals and Fisheries' (*Nature*, 1964, 203 : 116–18) Dr F. B. Worthington of the Nature Conservancy defended the cull as part of the modern philosophy of controlling populations. Man, animal, and plant do not live alone. They co-exist in a common environment. The explosion of one population can have good or ill effects on another. As has been frequently pointed out in this book, man's population expansion is crowding out many forms of wild life and must be halted if the environment for himself, his children and other species is not to be further debased and destroyed. We have rightly created sanctuaries and national parks for the benefit and conservation of living things and the preservation of the pristine landscape. Now we must accept the necessity to control populations of large animals as we have long accepted the need to control weeds and animals deleterious to agricultural production.

Worthington cites instances of protection within a nature reserve which have resulted in an increase of numbers to the detriment of the species itself, as well as of other species, including man: as deer and wapiti in North American parks; red deer on the island of Rhum, Inner Hebrides; and hippopotami and elephants in Africa which have partly destroyed their own feeding grounds. In New Zealand I saw how the red deer are killing out the young native forest trees in the South Island Alps and elsewhere and causing thereby erosion of steep slopes; introduced for sporting and food purposes, they now have to be shot as vermin, and usually the carcases have no value commercially because they lie in roadless country normally difficult of access. This is often true of other animals so controlled, where transport of the bodies is impossible: as with seals shot on the ice or upon remote islands – only the skin may be collected, the rest is waste. But at sites near roads or where transport can be found the harvesting of these surplus populations of deer, elephants, hippos and seals is proceeding more and more efficiently, as modern techniques are applied to ensure the maximum use of these products in the form of hide, horn, bone and meat. In a world which is already short of protein their economic use becomes more and more essential. In the case of British seals there is a small but valuable yield of hide, meat, bone meal and oil awaiting fuller utilisation.

It is not unlikely that, if the human population continues to increase, and the search for protein is developed by intensive fish-farming of the sea, the control of seal populations will become more severe in future. So long as such control is accompanied by careful conservation of stocks it may have a beneficial effect on fertility, which as we have seen, increases with hunting (elephant seal, and also the over-hunted harp seal).

The policy of controlling the numbers of grey seals during the close season, and upon islands which are under the management of the National Trust has had a stormy reception, but now seems to have been accepted by much responsible public opinion: Parliament agreed without much debate; perhaps the newspapers made too great a sensation of the killing of the baby seals which have such appealingly human faces, and after all the killing was humanely done by experts. Conservationists agree that there is no danger at present that the action so far taken under the culling policy will seriously reduce the numbers of the grey seal in its main breeding grounds, the British Isles, since it is applied only to one or two selected sites. The consequences of overcrowding, with no human control, in the main rookeries of North Rona and the Farnes, leading to a reduction of breeding success we have already discussed; and it may be desirable to cull at North Rona one day, if the grey seals continue to multiply there. Outside such densely populated reserves we have seen that the fishermen, when seriously troubled by the predation of seals, keep them in balance of numbers by shooting them in the open season, which in the case of the common seal is all the year round.

11 Distribution and Numbers

Whether the grey seal was originally an ice breeder, and became a land breeder or vice versa it is difficult to say. But it is interesting to find that it has adapted itself, in its several breeding grounds, to the moment when optimal success for its pups is most likely. On the stormy west and south-west coasts the main pupping occurs in September; in Scotland it is early October; but in the Farne Islands pupping is delayed until November although a few pups are born in the last half of October. This spreading of the breeding season into regional periods, so to speak, is very remarkable biologically, and difficult to explain where it occurs in the comparatively small area of British coasts, where regional climatic factors are not markedly different.

It may well be argued that being born in the autumn, with equinoctial gales and winter storms to face, is not the best time of year, and why are not grey seals born, like the common seals, in the calms of midsummer? The only answer would appear to be that the grey seal pup, dependent on the land for its survival during the first month of its life, during lactation and moult, is less likely to be hunted and exterminated by man in autumn and early winter than in the summer; and that disturbance by man over the summer months has resulted in an adaptation to breeding in the stormier seasons of autumn and winter, when inshore fishermen, their principal predator, are unable to hunt the nursery sites owing to adverse weather.

If we examine mean weather records with dates of inshore fishery activity together we find this is approximately true for all British coasts where seals breed. At the Farnes the inshore fishermen of that north-east coast are out later in the autumn; there is no heavy Atlantic swell from the west, but only short sharp seas which quickly subside, and as long as the weather holds good they fish around the islands. In the early days, before the general use of motors in craft enabled men to fish in winter more safely, before the Farne Islands began to receive 100 years of protection, there was, as we know, much persecution of seals there, and this hunting would take place normally as the seals came ashore to breed in early autumn. It may be, therefore, that the only pups reared in most years in those early centuries were those born after the sailing boats of the fishermen were hauled up for the winter, in late October and November.

In Canada and the Baltic breeding is still later. The pups are not born until the winter ice is formed and solid enough to provide a nursery, and no small boats can move about; or it breeds on islands rendered inaccessible by winter storms.

Recent investigations by Swedish and British naturalists has shown that about 2,000 grey seal pups are born in the Baltic from about 5,000 older seals which inhabit this sea, chiefly the central and northern waters. They are hunted by motor boat in the summer, and more intensively over the ice in late winter by Swedes, Finns and Estonians Bounties are paid to encourage the extermination of these 'competitors of the salmon, cod, herring and whitefish fishermen of the Baltic'. Probably one-quarter of the total population is killed annually, including many pups taken on the ice, so that it is considered that the grey seal population of the Baltic is dwindling.

The grey seal gives birth to its pup on the drift and fast ice in the Baltic, wherever this is thickest – in the Gulf of Bothnia and the central sea near and around the main islands of Aland, Gotland, Ösel and Dagö, and the innumerable islets near these. Pupping takes place from mid-February to April. Oliver Hook, who has visited this area, considers that due to the differing ice conditions from year to year, both as regards thickness and degree of drifting of ice by wind and current, there can be no regular annual rhythm of breeding, movement and distribution of the grey seals, and that the sites of pupping are dictated

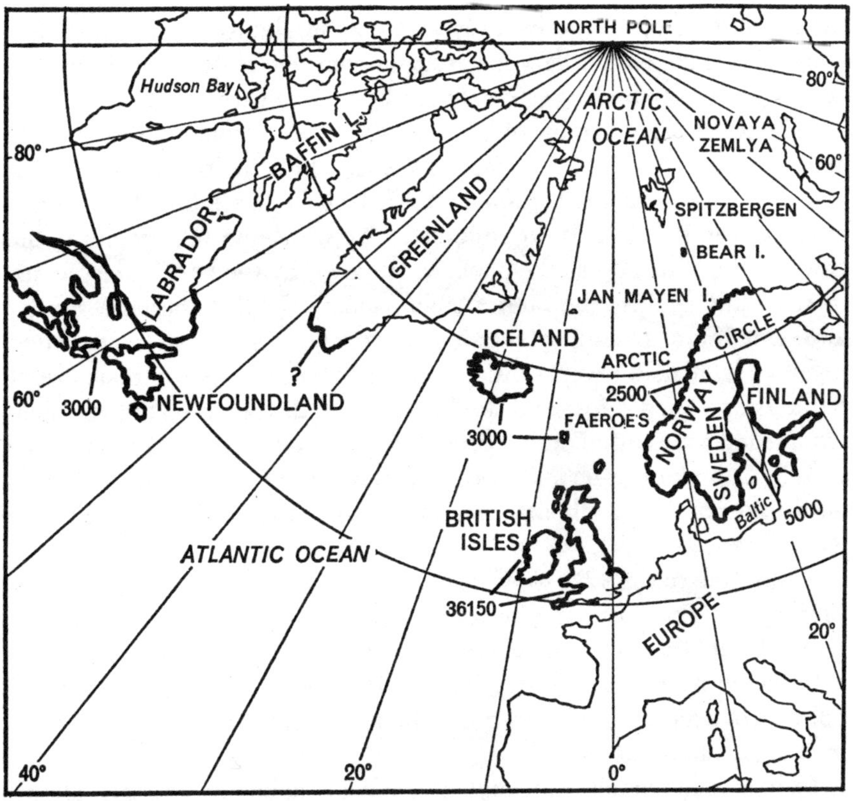

Figure 6. The thick line marks the limit of present world distribution (coasts frequented) of the grey seal. The figures are population estimates for 1965.

by the ice at the time of parturition. Fishermen who land on the ice in search of pups do not always find them in the same area as the year before. He adds boldly that 'It seems likely that gravid cows are able to delay parturition pending access to suitable breeding ice.'

Winter ice is the best, although not complete, protection for this seal in these overfished waters. In the usually ice-free south Baltic and the Kattegat area of Swedish-Danish waters it has become rare, though formerly breeding on Anholt Island.

The presence of the grey seal as a breeder in the Gulf of St Lawrence has been known for many years, but was not officially confirmed until the submission of distinctive grey seal jaws in claims for the ten-dollar bounty paid by local fishery interests for common seals killed. Formerly snouts were accepted as evidence for bounty rewards, until this fraud was detected. Two groups of grey seals were subsequently discovered: one breeding under the influence of the ice, the other on land. Pupping is in the coldest months, January to March. The ice-breeders behave much as the Baltic seals do, influenced by wind and temperature: the cows haul out to pup as the pack ice is blown by the strong west winds in the Gulf against the land (Amet Island, Northumberland Strait; and Deadman Island, Magdalen Islands), and becomes cemented together in a field several miles wide. The land breeding seals are at Sable Island (300 pups) and the Basque Islands (200 pups) in the open sea outside the Gulf. Total is probably 3,000 adult and sub-adult grey seals in Canadian waters, all breeding at a time when they are safe from local fishermen whose boats are hauled out to escape the rigours of the winter sea.

From time to time the grey seal is seen southwards along the coast of Maine, and also north to Greenland. About 3,000 inhabit the coasts of Iceland and the Faeroes, where they are hunted without restriction; here the pups are born during October in caves and on isolated beaches difficult of access, again at a time when the boats of the inshore fishermen are hauled out.

In Norway, home of seal mythology, seals have been hunted almost to extinction. When I made an estimate in 1954, from information supplied by Norwegian observers, there were only about 400 grey seals left along the west coast of that country. Per Øynes now reports (1964) that since 1956 'an invasion of this species has taken place from the coasts of Great Britain where this species is very numerous owing to legal protection. The increase has been especially considerable in the central parts of its Norwegian habitat. Breeding takes place in October–November and the influx from Great Britain is distributed on the same places that were traditionally known as grey seal rookeries'. He adds that today 'the grey seal is not much persecuted by hunters and the population will probably increase. The total population is somewhere between two and three thousand individuals'. Greatest numbers are

Figure 7. World range of the common seal, *Phoca vitulina*. There are five subspecies, including *mellonae* (isolated in the freshwater Seal Lakes, North America).

around the Trondheim fiords, with small groups almost to the North Cape.

The common seal has a similar distribution in Norway, but extending farther south and north, from the district of Møre to Finmark. Of this species Per Øynes reports 'The harbour seal used to be the most numerous [of seals in Norway]. The high prices paid for its skin has led, since World War II, to a severe decimation of this species in Norway, especially northern Norway, where the population amounts to only 1,500 individuals. In one district only, i.e. Troms, an average of 400 used to be shot annually before 1939. Some harbour seal is still shot during otter hunting, and if no protective measures are taken the harbour seal will become extinct in the fauna of northern Norway. Further south

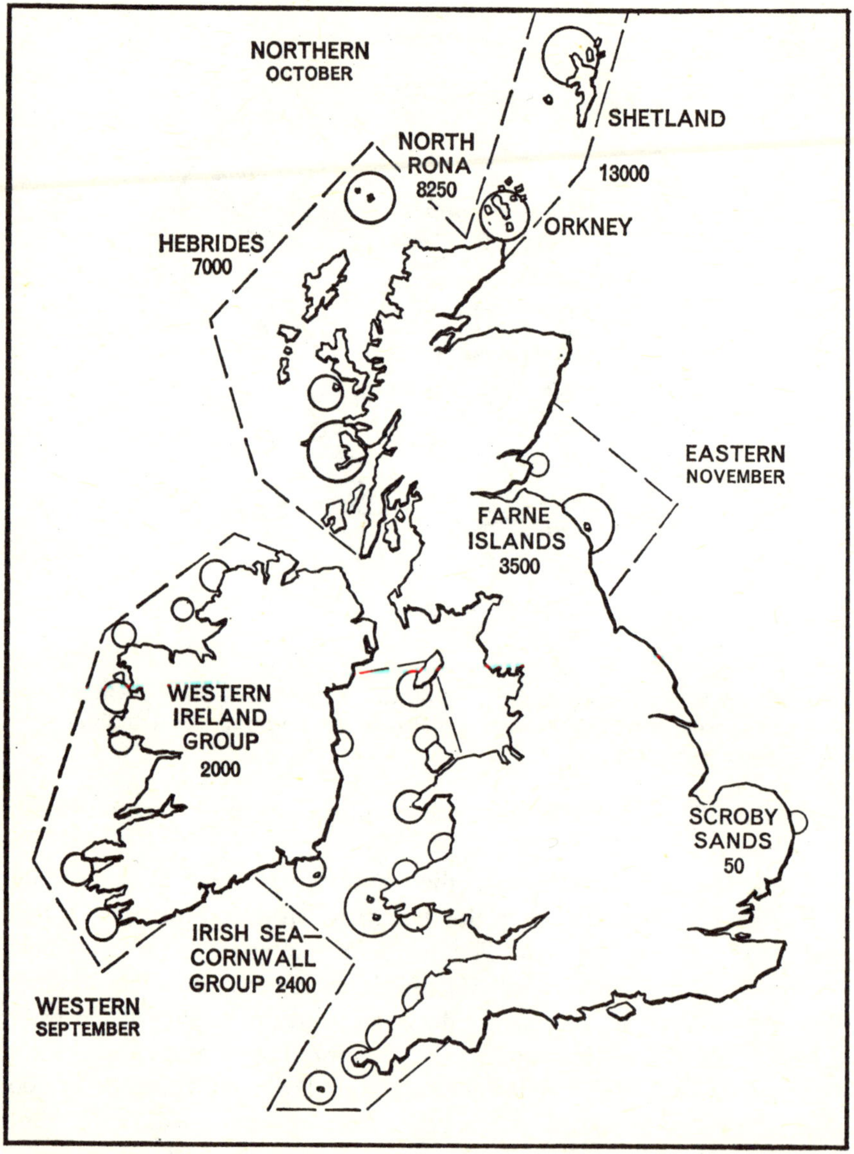

Figure 8. Main colonies of the grey seal in the British Isles, with times of calving on western, northern and eastern coasts.

it is more numerous. The district of Møre and Romsdal has the largest population, estimated to about 1,000 individuals. But even here the population is greatly reduced. The reduction of the population has thrown into relief its stationary way of life. It may be completely exterminated in one part of a fjord, whereas it is still numerous in other parts. It is, as a whole, no longer a real nuisance to the fisheries. It is piscivorous, preferring the species that are easiest to catch wherever it lives.'

The sketch map (Figure 8) shows the main breeding colonies of the grey seal in the British Isles. All coasts are frequented by the common seal, save those of southern and western England, and Wales, where it is only a sporadic visitor (although it has bred in the Severn estuary).

Along the whole coast of Ireland, which we have surveyed for the Nature Conservancy in autumn visits, both species of seal exist in small numbers, despite much shooting for bounty. The common is more numerous off the shallow east and north-east coasts but inhabits all the large sea-loughs and river estuaries of the west and south. On the east coast the grey is confined to Lambay Island (a small herd of about twenty-five individuals) where it breeds in caves, but is more numerous than the common seal on the south, west and north-west coasts, breeding chiefly on remote island beaches and in caves under high cliffs. Our survey, counts and estimates indicate minimum figures of 2,000 grey and 1,250 common seals for the whole of Ireland.

Table 3 does not include very small colonies, often of non-breeders, probably outliers from the main colonies, such as Lundy Island, Isle of Man, Bardsey Island, Scroby Sands (Norfolk) and Isle of May (Firth of Forth) where from time to time pup grey seals have been found. Numbers fluctuate at these small colonies, and are hardly worth giving here. Ringing has shown that young seals wander far (Figure 9) and are reported at some of these small assemblies. Seals so marked in Pembrokeshire have been reported a few weeks later from Brittany where small numbers lie out on rocks off the Ile d'Ouessant (Ushant), but no pups have been found there as yet.

Table 3 shows that approximately three-quarters of the world population of the grey seal inhabits British waters, and of these four-fifths

TABLE 3
Minimum Total Grey Seal World Population
(individuals above six months old)

		Pupping period
Cornwall and Isles of Scilly	400	September–October
West Wales	2,000	”
Ireland	2,000	”
W. and N.W. Scotland	7,000	”
North Rona	8,250	October
Shetland and Orkney	13,000	”
Farne Islands	3,500	November
Total British Isles	36,150	
Faroes and Iceland	3,000	October
Norway	2,500	”
Baltic	5,000	February–March
Canada	3,000	January–March
Approximate world total	49,650	

breed in Scottish waters. A world total of under 50,000 is not large for a species, adequate for survival, but without a substantial margin against natural decline or man-made catastrophe. In fact, it is, by numbers, one of the rarest seals in the world. The other rare seals whose numbers have been estimated are (not counting subspecies, which are often doubtfully valid): Ross seal 35,000, Hawaiian monk seal 1,500, Mediterranean monk seal 2,500, ribbon seal 25,000, Lake Saimen ringed seal 3,500, Lake Ladoga ringed seal 7,500, Baltic ringed seal 35,000 – but the Saimen and Ladoga ringed seals, like the race of common seals known as *mellonae* inhabiting the Seal Lakes of Ungava, Quebec, are fresh water lake seals, with no room to expand geographically as the sea-living species have. One seal is thought to be extinct or nearly so: the Caribbean monk seal, last seen near Jamaica in 1952, and heavily persecuted ever since Columbus killed seals near the island of Haiti in 1494.

Once again we emphasise the special responsibility of British people and British conservationists to protect such a fine and rare mammal as the grey seal which has its main breeding grounds in our islands, and is the largest and heaviest truly wild animal of our land-breeding fauna.

The total population of common seals is much less easy to compute, since it is impossible to count pups accurately, owing to their taking to the water so early after birth. But from all sources we have gathered the following rough estimates:

TABLE 4

Common Seals of the world – minimum estimates (excluding pups)

Phoca vitulina vitulina		
England (chiefly Wash area)	2,250	
E. and W. coasts of Scotland	2,000	
Shetlands and Orkneys	2,500	
Ireland	1,250	
Total British Isles	8,000	
Coasts of Holland	4,000	
" Germany	2,000	
" Denmark	500	
" Norway	3,000	
" Iceland	12,000	
Total Rest of Europe	21,500	
Phoca vitulina concolor		
Maritime Provinces of Canada	5,000	
N.E. Arctic Canada and Greenland	35,000	
Total Western Atlantic	40,000	
Ungava (freshwater) seal		
Phoca vitulina mellonae	500	
Total Atlantic population	70,000	
Pacific common seal		
Phoca vitulina richardi	50,000	(N.W. coast of America, California to Bering Sea)
Phoca vitulina largha	20,000	(Bering Strait to Japan and China)
Total Pacific population	70,000	

Minimum world population of *Phoca vitulina* is about 140,000 individuals. This is a conservative estimate; the real total could be double, perhaps treble. In his book *Seals, Sea Lions and Walruses* (1958) Scheffer estimates a minimum of 150,000 and a maximum of 451,000 *Phoca vitulina* of all subspecies.

The division of *Phoca vitulina* into five subspecies, each with its distinct geographical region, is doubtfully valid. Only the subspecies *mellonae*, isolated in the freshwater Upper and Lower Seal Lakes of the Ungava peninsula about ninety miles east of Richmond Gulf, Hudson's Bay, seems radically distinct, differing from its maritime neighbour *concolor* by its darker coat and slender coronoid process. These lakes communicate with each other and are 240 metres above sea-level, and frozen hard in winter. The small colony of seals may have originated during a post-glacial period of 3,000 to 8,000 years' duration, and has remained trapped and unable to reach the sea again. The north-western Pacific form, *largha*, is said to be bigger than the other subspecies, and to pup earlier – on the ice from February to April – and it may be born more often with the white woolly coat unmoulted.

Figure 9 shows the results of marking grey seals at British nurseries. Almost all the recoveries indicate, as mentioned earlier, that in the first few months of their life the moulter pups wander far, probably farther than they will ever do later in life. There is an explosion outwards from the nurseries in all directions, typical of most species of animate life.

Unfortunately few returns of marked seals occur after the first six months. This is almost certainly due to two factors: one, the disappearance of the rings or metal tabs, many of which are evidently lost within a year or two of being attached, either through salt erosion or other causes; secondly, few older seals compared with pups are recovered by shooting, netting or other means of capture alive or dead by man. Adult seals dead from natural causes are rarely met with. In forty years I have only found half a dozen dead adults in a condition to recognise any brand or metal tag, and none was so marked. I once found an aged bull dead in a cave, with no marks of violence, but more often

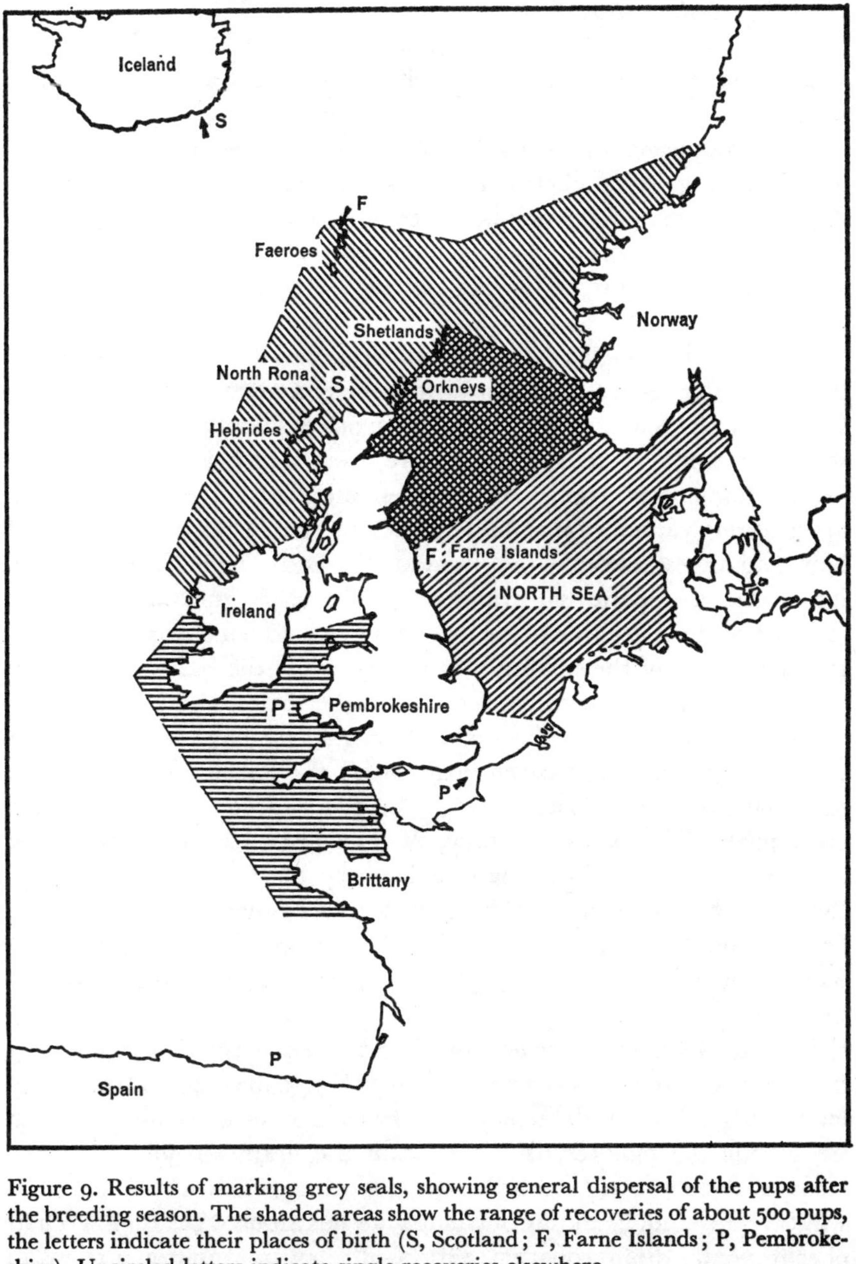

Figure 9. Results of marking grey seals, showing general dispersal of the pups after the breeding season. The shaded areas show the range of recoveries of about 500 pups, the letters indicate their places of birth (S, Scotland; F, Farne Islands; P, Pembrokeshire). Uncircled letters indicate single recoveries elsewhere.

I have found the body of a cow which has died during labour at a nursery.

The most permanent method of marking so far devised has been hot-iron branding; a cumbersome process not easy to carry out in the conditions of cave and small beach colonies under high cliffs, and best left to operators working in large colonies on open 'inland' sites such as North Rona and some Orkney islands. Here branding has had some success.

The first branding experiment with seals carried out in Britain was that made by the West Wales Field Society at Ramsey Island, Pembrokeshire on October 26th, 1946. Two irons, fashioned by a blacksmith in the form of a W, four inches square, were kept heated in turn on a primus lamp, and applied to eighteen moulted calves. The area of application (varied – on shoulder, back or rump) was first clipped of hair, and anaesthetised with a local anaesthetic. The business of carrying so much apparatus up and down the cliffs from one beach to another, and the keeping of the irons hot in a cold wind was exhausting, to say nothing of the holding of the pup while the hair-clipping and other preliminaries were carried out. Nevertheless the returns (over 10 per cent so far) have been rewarding. There have been two sight records thirteen and fourteen years later, the longest period for any seal marked in the British Isles. The first was of a bull, branded on October 26th, 1946, at Ramsey, observed fast asleep in fine condition on a rock 28 miles north-east of Ramsey on January 25th, 1959; the white brand W on the shoulder was very distinct. Another bull branded on that same day was seen again in perfect condition on Ramsey near where it was ringed – with white brand on his rump on October 13th, 1960.

Branding techniques are now much improved by the use of a marking iron kept at a white-hot temperature attached to the end of a blow-lamp fed by gas from a cylinder strapped to the back of the operator. One man sits astride the moulter, the other dabs the brand on for about two seconds. Even so calm conditions are desirable, which seldom occur on windswept island. Branding of young seals has aroused occasional bursts of sentimental disapprobation, expressed often in sensational terms in the popular Press. But in fact quick and permanent branding of one or two seconds' duration with a very hot iron can be done with almost no pain to the moulter pup, whose coat of blubber and thick hair seems remarkably insensitive. Cattle and domestic stock branding is, by comparison, much more disturbing. The disadvantages of branding on a large scale is that it is impossible to devise a long series of serial marks, identifying each individual and large enough to be recognised at a distance by the average observer (who will be reporting back to the marking organisation) without covering the animal with an inscription! Even the huge elephant seal, with its enormous area of blubber, has to be deeply cut or branded in large letters and figures to remain identifiable as an individual by man for life.

However, although various types of metal or plastic rings or tabs placed on the tail or on a digit of fore or hind flipper have the disadvantage of being lost within a few years, the returns in the first few

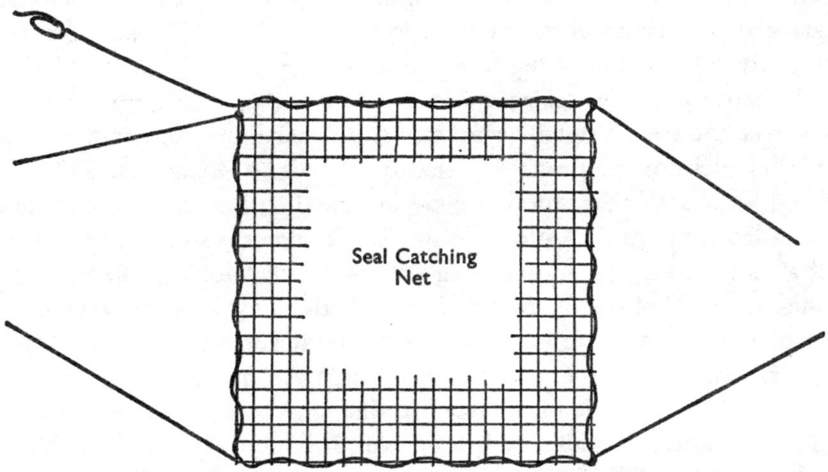

Figure 10. Details of the net used for catching adult seals. The four spreader ropes open and spread the net; the thick footrope runs through the outermost meshes. When a seal strikes the net this rope is pulled tight, thus completely bagging the seal. The net is ten feet square with a six-inch mesh made of quarter-inch nylon, and it will hold a bull weighing half a ton.

months have been valuable in ascertaining the movements of the independent pups (Map, page 159). Generally it has been found that monel (composed of nickel 67 per cent, copper 30 per cent, iron 1.4 per cent, manganese 1 per cent) tags are the longest lasting of metal rings. We have used these to mark some grey seals in Wales for over ten years. So far our longest recovery of a seal so marked is that of a female ringed when ten days old on Ramsey Island on October 2nd, 1955, which was caught a year later in the same place; it was next reported from near Truro on November 20th, 1960, at five years of age.

Ringing in British colonies with perishable rings or tags has largely served its purpose of tracing the early migrations of the pups, and until some more lasting form of easily applied marking is invented, marking has been discontinued in the smaller colonies. It is known that fur seals, handled and tagged or branded by man, are disturbed to the extent that they grow rather more slowly afterwards; this is proved by weighing tagged and untagged pups. For the purpose of examining wild seals alive it is now possible to immobilize these temporarily with darts charged with tranquillizing drugs, so long as they can be kept on land until the effects of the drug have worn off.

In capturing older seals we used a home-made net (Figure 10) about ten feet square, of nylon (synthetic fibre) cord (not too thin, as this might cut into seal flesh; we used quarter-inch diameter). This was fitted with a thicker running rope attached to one corner only, and threaded through the outermost meshes. The net is spread in the path of a seal, two operators standing one each side, holding the spreader lines, and one of the operators also holds the loose end of the running rope. The net can be moved, as spread, to intercept the seal as it heads for the sea, and as soon as the seal crosses the middle of the net, the operators flip the side of the net it is facing over its head with the aid of the spreaders. Usually the head passes through the six-inch mesh and holds at the seal's 'shoulders'. The running rope is immediately pulled hard, and the spreaders dropped; and this encloses the seal as a banana in a string bag. The running rope now has a free length of over forty feet, and can be wound round a convenient boulder to anchor the netted seal; or two men can easily hold even a half-ton bull at the end of the rope, since the seal's flippers are pressed against its body and it has almost no purchase to move forward with strength. To set the seal free after examination and marking, etc., it is only necessary to drop the running rope loose and pull on the spreader ropes firmly and steadily. This opens the nets automatically and it can be drawn over the seal's head, and once its hind flippers are freed, the seal slips out backwards.

The pursuit, capture and marking of animals for the better understanding of their life histories satisfies two main natural urges in man. I admit to very great pleasure in hunting seals in this way; this pleasure springs from a scientific curiosity to find out more about seals without harming them, and from the joy of the chase in such marvellous surroundings of clean sea, wild cliff, lonely island and dark cave, a chase in which the cunning of the hand with oar, sail and seal net is pitted against the hazards of rough water and weather, steep cliff, boulder, beach and cave, and the keen intelligence of the quarry. There are many expected risks and some unexpected adventures to be enjoyed in seal marking. More than once the tiny dinghy used for entering caves has capsized due to a heavier surf than expected which brought in a wall of white water, and threw me and my helpers into the sea; or a party of seals has rushed out beneath the boat in water so shallow that we have been thrown into the sea with them. Fortunately seals do not bite humans unless cornered or captured; we have only been bitten as a result of our own negligence in approaching a seal. On one occasion a bull seal in a net tore my trousers down – luckily only grasping a mouthfull of corduroy – to the great joy of my helpers. On another occasion I slipped accidentally into a pool in which a cow was hiding with her calf; there was no room for the three of us, and before I could get out the cow caught me by the leg – and held on long enough to leave a permanent scar.

There is the considerable excitement of stalking over slippery low water rocks, of laying the net in the right place to intercept the seal; of guiding it into the net by judicious appearances from behind cover. In a cave known to harbour sleeping seals (often detected by scent) one man must crawl in and coax the seals out; often the passage is narrow and it is a race to get out before you are overtaken by the panicking seals. They may charge in a mass, and the netsman must decide how many he will try to catch – one is usually enough if it is adult, two in a net will usually fight with each other.

Often we have rescued young seals in difficulties, which would otherwise have died miserably. Some we have found wedged between boulders, their hind flippers suspended in space, or otherwise caught in a hole in the rocks by movement of the sea at high tide tumbling the boulders about. Once we rescued a young man trapped on a narrow beach; he had sprained his ankle in climbing, and was unaware that in a few more hours he would have been driven into a cave with small chance of finding a ledge above high tide.

A curious rescue took place at the back of Ramsey Island. We had landed on a narrow pebble beach shut in by unclimbable cliffs. A huge cow seal was lying beside her new-born white pup. She refused to move or allow us to touch her calf – unusual but not unprecedented behaviour. But soon we saw why. Out of the boulders above us came a bedraggled little corgi dog, wet, thin and out of coat. He whined and wagged his stump of a tail frantically, and leaped into the arms of my young assistant, Joanna. He stank of the high body odour of the seals he had been living with. Probably he had fallen into the sea while chasing cliff-rabbits, and had swum to this rock-walled dungeon enclosing the nursery of the grey seals. Six pups had been born there in the past few weeks, and we marked all these with rings. Somehow the corgi had survived a month or more by eating the afterbirths of these pups – he vomited a mass of placental flesh which proved this. He had had nothing else to eat, because the cows would have savaged him if he had approached their pups. But for our casual visit he would have perished in the gales which subsequently drove rough seas to fill cave and beach over the high equinoctial tides. We put Timmy the corgi aboard the motor boat and sailed eastwards to Ramsey Island harbour. The Davies family who were then farming the island welcomed him as one long mourned dead; Timmy had been lost forty days and forty nights – in the seal fastness. He was washed, groomed, scented liberally and petted no end. But after we had drunk coffee with the grateful family Timmy followed us to the boat. He smelled at my trousers, well-flavoured with seal scent, lingeringly. We had to leave him, whining in restraining arms. The perils and adventures of the seal life had gone to his head.

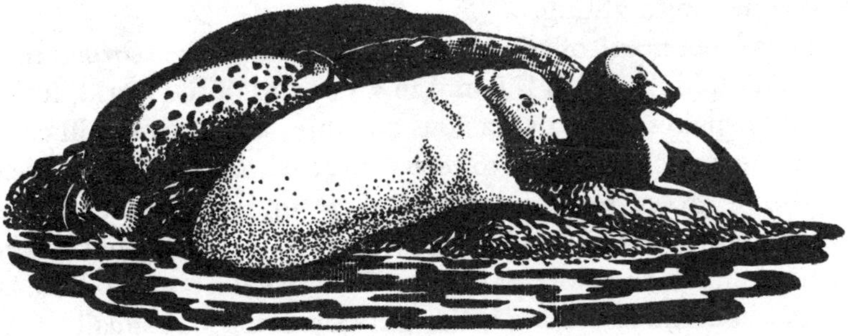

12 Predators and Mortality

In British waters seals have almost no enemies except themselves and man; and normal diseases which cause mortality. Their relations with man have been already discussed; as well as their own social stresses of overcrowding and fighting.

Much as been made of the fact that the killer whale (*Orcinus orca*; it is really a dolphin), which is found in all oceans of the world, feeds on seals and porpoises. It is a voracious creature, and several will concert together like a pack of wolves to tear a large whale to pieces: a meal of half a dozen grown seals is commonplace for this animal – thirteen porpoises and fourteen seals in one stomach is an oft-quoted record. I have seen a small school of two or three killer whales a mile off the large seal colony on Ramsey Island, the tall dorsal fin majestically cutting the tide-rip southwards – this party appeared to be migrating, not feeding. I have also watched that same fin ripping the smooth surface within the deep water of Milford Haven, as a pair of killer whales glided there. But I have never seen any evidence locally of direct attack on seals close inshore, where the killer might, one supposes, fear to hunt in case of stranding.

No doubt the majority of grey pups, migrating between France and Ireland, or over the North Sea (as marking them has proved) would stand small chance of surviving an encounter at sea with killer whales. In one late autumn there was quite an epidemic of headless grey seal pups and one or two adults washed up on Welsh shores (and one in Northumberland); and it was thought that the playful killer whales, satiated perhaps with whole carcases, had nipped off these heads as a *bonne-bouche* to their seal supper. Gavin Maxwell, in *Ring of Bright Water*, quotes a probable killer whale as responsible for the crushed head and gaping flank wound of a newly dead brown seal washed ashore near his house in the Hebrides, and he also describes how the lighthouse men of Hyskeir have seen 'killers slash seals for sport and not for food, and leave them maimed and dying among the skerries. A single bull killer terrorised the little harbour of the Isle of Canna one summer'.

It is strange that the killer whale does not multiply and terrorise more of the smaller sea mammals on which it preys so ferociously, for it appears to have no natural enemies to limit its increase. Yet although it ranges British waters regularly, and is even – inexplicably – sometimes stranded alive on the shore, it never appears in large numbers, and is very rarely seen to attack seals there. This may be because the attacks take place at sea; but generally it is considered that this predator has no observable effect on British seal populations. Anyone who has watched the contortions of a seal at play (for example, in an aquarium tank) and seen how it can double itself into a perfect U, swim upside down, see clearly with binocular vision at every angle under water, will realise that even a killer whale might have difficulty in catching a seal wise enough to adopt evasive tactics rather than try to outswim the killer – which it could not do.

Large sharks are considered to be predators of the smaller seals, although I can trace no sight records of such attacks on British seals. Perhaps a few wandering pups are eaten by sharks. In the Galapagos Islands sharks are the worst enemies of the seals; their stomachs usually contain the remains of young seals. In the Arctic walruses occasionally kill seals and eat them, though ill-equipped for taking meat food; and of course polar bears live much on seals caught on or near icefields. Possibly the white-tailed or sea-eagle, rare in the British Isles, takes an undefended young pup at times; but there is no record of the golden eagle doing so, although it will kill quite large deer fawns.

Seals do not normally suffer from epidemic diseases. Those which breed on ice, in frozen latitudes, are born under almost sterile conditions, the low temperature is unsuitable for the multiplication of bacteria externally, and the only infection would come from the mother or from the faeces of visiting sea-birds, especially gulls feeding on the afterbirth and seal droppings. But in temperate latitudes, such as the British Isles, pups are born on beaches and soil often contaminated with the faeces of sea-birds, of other seals, with rotting seaweeds, dirt and debris. If these pups receive cuts and wounds from sharp rocks and the manœuvres and bites of adults at the nurseries, sepsis is common and can be serious.

On October 28th, 1963, I landed on the Little Skerry, an isolated seldom-visited islet in the furious tide-race of the Pentland Firth, to find some hundred grey seal pups scattered over the muddy surface of this tiny island which in summer is packed with nesting sea-birds. Practically all the vegetation had been sloughed flat by the passage of the adult seals, and with the salt spray and rain the soil was a sticky porridge of dirt and faeces. The eyes of many of the pups were in a septic condition, purulent and closed, and many had died, probably from conjunctivitis and associated blindness.

Blindness in seals is not uncommon, but does not necessarily result in death. I have seen several healthy, totally blind grey seals. On this occasion at Little Skerry one mature cow did not move from beside her calf on our arrival, although most of the other adults plunged away down the slimy slopes to the sea. Her eyes were opaque. She was quite blind, and although she could hear and scent us, as we stood near her, she only expressed her uneasiness by patting her calf, a healthy fat white child, with one fore-flipper, thus reassuring herself of its presence. We left her in peace, admiring her ability to thrive in blindness: in this condition she had made the climb up to the top of the islet where we found her, over rough rock ledges, possibly remembered by feel from previous years. She was in excellent fat condition, and had obviously lived a normal life, fishing successfully, breeding, etc., despite total blindness.

On another occasion, when catching adult seals in nets in Wales, for marking purposes, a friend and I patiently stalked a huge bull seal asleep on the Red Wilderness beach one winter's day. The tide was well out and the bull rested on the pebbles above the high water mark. We wriggled low on our tummies over the wet boulders along the edge of the sea and silently, slowly got between the bull and the water without his seeing us. We need not have troubled. Although he woke up and became uneasy as we stood up and spread the net to intercept him as he would plunge seawards, he could not see us. He sniffed the air for our scent, but did not attempt to move until the net was closed around him; and by then we had realised he was blind, from his open opaque eyes and his behaviour. He snarled and snapped at the meshes, trying to break loose. He was in normal fat condition. On release from the net after we had ring-marked him, he bounced downhill into the sea. Half an hour later we saw him return cautiously, feeling his way through the heavy surf, blind but determined to resume his siesta on the smooth pebbles.

Other seals have been found blind in one eye; and many older seals also carry septic cuts and scars, as well as fresh wounds, presumably the result of fighting with each other or with large fish prey, or possibly through swimming ashore through sharp rocks. Cow seals inflict small bites on their own pups in trying to pull them back from a rough sea; we frequently find the 'ankles' of young pups septic from teeth-marks made by the cow in grasping a pup in this way.

Fleming, in *A History of British Animals* (1828), records that 'about fifty years ago, multitudes of carcases [of seals] were cast ashore in every bay in the north of Scotland, Orkney & Zetland, and numbers were found at sea in a sickly state'. What the cause of this sickness was we do not know. A morbidity of grey seal pups on the rocky shelves of St Kilda is reported by Gallacher and Waters, who state that sixteen out of forty-five pups there in 1960 died of pneumonia, probably through premature exposure to sea in the rough terrain unsuitable as a seal nursery. Three other pups of moulter age were in the early stages of respiratory infection caused by pneumococcus, and two also had conjunctivitis, due probably to rubbing of the eyes with virulent pus from the nose when scratching. This trio were treated with penicillin, and recovered completely. Antibiotics are now used extensively in the treatment of respiratory and other diseases of seals in captivity, and with much success.

We have seen that seals harbour the cod-worm, which may pass undetected during the processing of cod, as well as other fish, for human consumption, and is harmless to man if the fish is adequately cooked. Mohr (1952) gives a list of fifty parasites for seals and walrus in European waters, including round-worms, tape-worms, flukes, thorn-headed worms, nasal mites, and lice. It was discovered, by Olsen in 1962, that the hookworm parasite, causing considerable mortality in the Alaskan fur seal, is transmittable in larval form through the milk, from the nursing cow to its new-born calf.

Yet in general it is rare to find a sick adult seal. Except when moulting – or nursing – seals are fat, covered with fur and blubber resistant to the cool weather and rough seas of the latitudes they inhabit.

They are, I think, a happy people. In the words of L. A. G. Strong, they are:

The old grey music doctors
Of the ocean
Their holy, happy eyes
Shining devotion.

Ballad of the Grey Selchie of Sule Skerrie

(There are many versions of this ancient ballad in the Shetlands, Orkneys and Hebrides. The following version is a mixture from the best of these, preserving both the story and the melodic scansion. It is interesting to realise that, from the results of marking grey seals, Orkney selchies do cross to 'Norway land'.)

In Norway land there lived a maid;
'Hush loo lillie,' this maiden sings,
'I know not where thy father is –
By land or sea his travellin's.'

It happened on a certain day,
When this fair maid fell fast asleep,
That to her came a grey selchie
And set him down at her bed feet.

Said he, 'Awake, my pretty maid,
For oh, too soundly thou dost sleep!
I'll tell thee where his father is –
He's sittin' close at thy bed feet!'

'I pray thee tell to me thy name,
Oh tell me where thy dwellin' be?'
'Ah, I've come far, frae west of Hoy;
I earn my livin' in the sea.

'I am a man upo' the land,
I am a selchie in the sea,
An' when I'm far frae every strand
My dwellin' is in Sule Skerrie.'

'Alas, alas, this woeful fate,
This weary fate's been laid for me –
A selchie frae the west of Hoy
To Norway land to mate wi' me!'

'My dear, I'll wed thee wi' a ring,
Wi' weddin' ring I'll marry thee.'
'Thou go thee weddin' where thou wilt,
Be sure thou'll ne'er wed none wi' me.'

'My wee young son thou'll nurse for me
For seven long years upo' thy knee,
An' at the end of seven long years
I'll pay thee full the nourrice fee.'

Now when he brought a purse of gold,
And when he'd put it on her knee,
Said he, 'Gi' me my seven year old
And take thee up thy nourrice fee.'

'Alas, alas, this woeful fate,
This weary fate's been laid for me.'
And now aloud she sobbed and sighed
As if her heart would brake in three.

'Oh now I'll wed thee wi' a ring,
Wi' weddin' ring I'll marry thee!'
'Thou go thee weddin' where thou wilt,
Be sure thou'll ne'er wed none wi' me!

'I'll put a chain around his neck,
A gey good golden chain 'twill be;
If ever he comes to Norway land
Thou'll have a gey good guess on he.

'An' thou wilt wed a gunner good,
A gey good gunner he will be,
An' he'll go out on May mornin',
An' shoot the son of grey selchie.

Oh, she has wed a gunner good;
A gey good gunner it was he
Who went him out on May mornin'
And shot the son of grey selchie.

He took the chain from round the neck
Of dying son of grey selchie –
A sea-bright golden chain it was –
And called his Norway wife to see.

'Thou's shot him dead, my little son,
Who lived wi' selchies in the sea,
Oh, far he went frae Norway land,
His father dwells in Sule Skerrie!

'Alas, alas, this woeful fate,
This weary fate's been laid for me!'
And loud and long she sobbed and sighed:
Her tender heart it brake in three.

Index of Subjects

Adolescence, 54–60
Age at breeding, 59, 120
Aged seals, 30, 70
Assembly, 54, 66

Ballad of Sule Skerrie, 131, 171
Baltic, 67, 150
Battue, 131
Basking, 61, 65
Beachmaster, *see* Bull, master
Birth, *see* Parturition
Bladdernose seal, 78
Blastocyst, 61
Blindness, 29, 81, 168
Blood, 80
Blubber, 49, 61
Body temperature, 80
Bounty for killing, 140, 150
Branding, 160
Breeding dates, 149
Breeding season, 85–113
Brittany, 49, 112, 126
Bull, bachelor, 96, 108
Bull, behaviour of, 35, 43, 70, 86, 100
Bull, challenger, 86, 94, 107
Bull, master, 86, 106

Calf, *see* Pup
Canada, 40, 67, 83, 129, 144, 150
Claws, 75
Cod-worm, 143
Coition, *see* Mating
Colonial breeding, 124
Common seal, adult, description of, 16
Common seal, distribution, 153
Common seal, fresh water species, 158
Common seal, habitat, 67, 114
Common seal pup, behaviour, 36, 114
Common seal pup, captive, 20
Common seal pup, description, 41
Common seal pup suckling, 38
Control of seals, 127, 145
Cornwall, 111, 126, 137
Culling, 145

Dances, 58, 72, 91, 115
Diseases, 167
Disturbance, human, 131
Distribution, 159–65
Diving, 78
Dolphin, 15, 19, 29, 32, 52, 72
Dominance, *see* Hierarchy
Dugong, 29

Ear, 81
Eared seals, 16
Echo-location, *see* Sonar
Elephant seal, 32, 35, 52, 59, 92, 103, 119

Eskimo seal-hunter, 130
Eye, 15, 80

Faeroe Islands, 48, 64, 132
Farne Islands, 49, 51, 64, 96, 124, 130, 150
Fasting, 44, 143
Feeding rhythm, 65
Fertility, 59
Fighting, 101, 109
Food of adults, 52, 82, 143
Food of pups, 40, 51
Frisian Islands, 114
Fulmar petrel, 33
Fur of seals, 15, 66, 115
Fur seals, 16, 32, 41, 52, 59, 103, 119

Grassholm Island, 125
Green Holms, Orkney, 98, 106
Greenland, 130, 152
Grey seal, description of adult, 16
Grey seal, distribution, 149
Grey seal, habitat, 67
Grey seal pup, behaviour, 43, 49
Grey seal pup, captive, 24
Grey seal pup, description, 42
Grey Seals Protection Acts, 131, 141
Gulf of St. Lawrence, 144, 151

Harem, 43, 103, 110, 119
Harp seal, 52, 129, 144
Haul-out, *see* Assembly
Haskier, 132
Hearing, 30
Heart, sphincter muscle of, 79
Hebrides, 43, 105, 132
Hierarchy, 50, 69
Hunting, 40, 46, 115, 138, 145

Ice-breeding seals, 129, 150
Iceland, 132
Implantation, delayed, 61
Individual distance, 57, 67, 69, 96
Inland, grey seals breeding, 95
Intelligence, 15, 21, 37

Ireland, 49, 111, 134
Isle of Man, 126, 137, 155

Killing, *see* Hunting

Lactation period, 39, 44
Legends, 29, 133
Locomotion on land, 74
Longevity, 120

Manatee, 29
Man and seals, 128–48
Marking with rings, 26, 66, 158
Maternal care, 35, 38, 42, 44, 111
Mating, common seal, 116
Mating, grey seal, 92, 116
Meat, seal, 132, 134
Mermaid, merman, 29, 68, 135
Migration, 47, 52
Monk seal, 64, 156
Mortality of adults, 119, 166
Mortality of pups, 38, 43, 51, 106, 125, 166
Moult, 41, 46, 53, 115
Moulter (pup), 46–51
Music, response to, 23

Netting of seals, 66, 162
Non-breeding adults, 54–60
North Rona, 51, 95, 125, 132, 138
Norway, 49, 129, 152

Odour of seals, 22, 66, 94
Oestrus, 91, 102, 110, 111
Oil, seal, 130, 134
Orkney Islands, 34, 43, 51, 83, 95, 106, 146
Oronsay Island, 105
Otter, 76
Overcrowding, 125

Parasites, 170
Parturition, common seal, 32
Parturition, delayed, 33, 151
Parturition, grey seal, 31
Pembrokshire, 33, 44, 51, 64, 138

Pentland Little Skerry, 95, 168
Playful behaviour, 39, 45, 52, 54, 72
Pneumonia, 169
Polygamy, polygyny, 60, 117
Population, 118–27
Predators, 128, 166
Pregnancy period, 61
Pribilof Islands, 51, 110, 119
Protection, 131, 140
Puberty, 59, 124
Pup, artificial feeding, 20
Pup, description, 34
Pup, natural feeding, 31, 36, 97
Pup, weaning, 39, 45

Ramsey Island, 33, 85, 105, 125, 160
Rare seal species, 156
Recovery of marked seals, 158
Red Wilderness, 54, 58, 70, 126, 168
Roan (Irish), 136
Ron Mor (Gaelic), 132
Rusk Holm, Orkney, 106

St. Kilda, 139, 169
Safety distance, 57, 67, 69
Salmon as seal food, 141
Sea-lion, 16, 24, 29, 70, 78
Sea monster, 136
'Selchie', 131
Sense of hearing, 30, 81
Sense of sight, 30, 80
Sense of smell, 30, 81
Sense of touch, 81
Septic condition, 168
Sex ratio, 34, 118
Shark, 112, 167
Shetland Islands, 38, 40, 48, 58, 114, 134, 138
Shillay Island, 95, 134
Shooting of seals, *see* Hunting
Singing, 23, 68, 137
Skeleton, 15
Skokholm Island, 48
Skomer Island, 85, 125
Sleep, 40, 65, 69, 77
Smalls Lighthouse, 66
Social behaviour, 69, 124
Sonar, 29
South Ronaldsay, 106
Spring mating, 55, 63, 115
Storm dangers, 44, 67
Sule Skerry, 131
Sulisgeir, 138
Superstitions, 68
Swimming, description of, 72, 77

Tameness, 28
'Tangfish', 114
Tears, 81
Teeth, 39, 82, 120
Territory of bull, 86, 105
Territory of cow, 43
Tide dangers, 38, 111
Twins, 34

Voice, 23, 30, 42, 68
Virgin, behaviour of, 54, 110
Vital statistics, 119–27

Wales, 48, 64, 96, 105, 111, 137
Walrus, 16, 167
Wash, the, 36, 40, 66, 83, 114
Weaning, 40, 45
Weddel seal, 79
Weight, 16, 35, 44, 52
Whale, 15, 29, 32, 52, 59, 80, 129
Whale, killer, 73, 112, 166
Whiskers, 81
Worms, *see* Parasites